ALL NEW ►Bathroom
IDEA BOOK

SANDRA S. SORIA

ALL
NEW

Bathroom

IDEA BOOK

The Taunton Press

To my boys, Ricky, Lucca, and Elijah

Text © 2009 by Sandra S. Soria
Illustrations © 2009 by The Taunton Press, Inc.

The Taunton Press
Inspiration for hands-on living®

The Taunton Press, Inc.
63 South Main Street, PO Box 5506
Newtown, CT 06470-5506
e-mail: tp@taunton.com

Editor: Erica Sanders-Foege
Copy Editor: Laurie Baird
Cover design: Kimberly Adis, Brooke Rane
Interior design: Kimberly Adis
Layout: David Giammattei
Illustrator: Jean Tuttle
Cover Photographers: Front cover: (left, top to bottom) photo © Olson Photographic, Design: Amazing Spaces, Briarcliff Manor, NY; photo © Eric Roth, Design: Heidi Pribell - www.heidipribell.com; photo © Tria Giovan; (middle) photo © Olson Photographic, Design: Studio DiBerardino, Darien, CT; (right, top to bottom) photo © Verity Welstead/Redcover.com; photo © David Duncan Livingston; photo © David Duncan Livingston; Back cover: (top left) photo © Bieke Claessens/Redcover.com; (bottom far left) photo © Tria Giovan; (middle) photo © Douglas E. Smith; (right) photo © Eric Roth

Library of Congress Cataloging-in-Publication Data

Soria, Sandra S.
  All new bathroom idea book / Sandra S. Soria.
     p. cm.
  ISBN 978-1-60085-086-8
  1. Bathrooms. 2. Interior decoration. I. Title.
  NK2117.B33S67 2009
  747.7'8--dc22

                    2009030693

Printed in the United States of America
10 9 8 7 6 5 4 3 2 1

The following manufacturers/names appearing in *All New Bathroom Idea Book* are trademarks:
American Standard® *(Compact Cadet®),* Bosch®, Dolce & Gabbana®, Dupont™ Corian®, ENERGY STAR®, EPA Water Sense®, evolve™, Formica®, Kohler® (DVT™, Forte®, Fountainhead™, Hatbox®, Power Lite™, Purist®, RiverBath®, San Raphael™), Nevamar®, Noritz®, Pionite®, Ralph Lauren™, ShowerStart™, Silestone®, Susan Sargent®, Tommy Bahama®, Toto®, Wilsonart®.

acknowledgments

THE VOICES, IDEAS, AND CREATIVITY OF MANY ARE CONTAINED WITHIN THESE COVERS.

It's difficult to know where to start, or end, when it comes to thanking those who have in some way contributed to this book's completion. I suppose I can't go wrong by thanking my editor, Erica Sanders-Foege, for her good humor and good ideas throughout this one-year process. Mostly though, I'd like to thank her for taking a chance on turning a magazine maker into a book author.

This book is filled to the rafters with examples of the best in home design today. In covering the shelter industry for 25 years, I'm always astounded at the seemingly never-ending amount of originality, ingenuity, and craftsmanship that is dedicated to creating a gracious and expressive home. I salute the talent of all the architects, interior designers, craftspeople, product designers, homeowners, photographers, stylists, and editors who in some way put a mark on this book.

There are many who have contributed precious knowledge and time to this project. Though I can't thank you all individually, I hope you feel your knowledge has been sifted well into this volume. I would like to call out the National Kitchen and Bath Association (NKBA), the National Association of Home Builders (NAHB), the National Association of the Remodeling Industry (NARI), and the Tile Council as wonderful clearinghouses of information that is useful to a homeowner intent on embarking on the detailed and decision-fraught process of building or remodeling a home.

Naturally, I'd like to thank my dearest ones for being my constants during a year of transition. My sons haven't flinched at seeing flitting corporate executive Mom become a more nested freelance writer Mom. And my darling husband hasn't yet run screaming from my more hovering presence. My mother must be thanked for always knowing how to put up with the moods of a writer.

contents

introduction

MY TWO TEENAGED SONS DIDN'T BELIEVE ME when I told them their Grandma (a spry 71-year-old who has been known to outlast them at certain games) grew up in a rural farmhouse without indoor plumbing. In her time, she has seen the bathroom come in from the backyard to take up three or four or more spots on a modern home's floor plan.

Bathrooms have not only come in from the cold, they have gone way past the utilitarian spaces of a few decades ago to being one of the most remodeled spaces in the home. These days, we value the bath as a personal, pampering space. For some of you busy homeowners, it's the only quiet time in the day! It's also a room that has a direct ability to affect the resale of our homes.

Though they are often the smallest spaces in the home, bathrooms garner enormous attention. Designing a bath can be a daunting process thanks to the technical nature of plumbing—and to the sheer number of choices available to the shopper.

Enter the *All New Bathroom Idea Book*. This volume breaks down the process of designing a bathroom into a logical sequence, from planning the space, to selecting a style the family can live with and love, to exploring the latest options in materials and fixtures.

Detailed captions and multiple sidebars will guide you to the information so you can realize the new bathroom of your dreams. All the elements of today's bath are explored, from basic faucets to Zen shower stalls.

Best of all, hundreds of images of inspiring new bathroom designs will allow you to tour the best baths we found around the country, gathering innovative and well-designed ideas as you go. You can use these spectacular photos to "shop" for decorating ideas you'll love, and to find creative and practical solutions for your own space. Or use the book to communicate with your carpenter, contractor, or design team, relying on the old adage that a picture is worth a thousand words.

In your hands is a virtual designer's notebook of ideas, inspiration, and information. This volume is a wonderful first step in turning a dream bath into a real one, whether it's a new construction or a remodel, a tiny attic bath or a luxurious master suite. It's all here for you. Well, except the new eco-outhouse concept. I'm not sure Grandma is ready for that one yet.

—Sandra S. Soria

getting started

• • •

SELECTING MATERIALS FOR A BATHROOM REDO USED TO BE ABOUT as exciting as deciding which brand of oatmeal to buy. Today, the range of design options includes all the colors, shapes, textures, and surprises of the whole cereal aisle! Bath fixtures have stepped out of the box flaunting shapely new silhouettes. Surface materials come in all textures and types—even nationalities—from slabs of natural stone to sleek sheets of marble veneer to bamboo.

Even the bathroom's function has expanded in wonderful ways. Though this essential room is still expected to meet the family's basic hygienic needs, it also doubles as a relaxing retreat. Little luxuries and everyday extras are available to make your bath personal and pampering, no matter what your budget.

These abundant choices make designing the best bath for your home more exciting—and more complex. Planning is essential. Where to begin? Picking up this book is a great start. Flipping through shelter magazines can help you get a fix on your favorite looks. You can also walk the showrooms of specialty bath design shops and home improvement centers to get a sense of your options. If you're an Internet user, it's time to put your Web search engine in high gear.

As you get a sense of what's in the market, consider your own needs. How has your bathroom fallen short of your needs in the past? Who will be using the new bath and how often? How much do you want to spend? This chapter will help you answer those questions and guide you as you make decisions about your space's look, layout, and level of luxury.

Bathroom fixtures with pared-down profiles and surfaces of polished, natural materials give modern baths serenity and style. This Victoria and Albert Napoli tub is low-slung and lightweight, making it a practical focal point in this sleek space.

dollars and sense

●●● YOU'VE LIKELY GOT THE SPACE SELECTED FOR YOUR NEW OR RENOVATED BATHROOM; now it's time to determine what should go into the space and where. For most of us, those elements are largely determined by budget. So let's talk dollars and sense.

Generally speaking, a minor makeover, such as a new toilet or sink and a fresh coat of paint, can be accomplished for under a couple thousand dollars. For a more substantial do-over, one in which you keep the layout but purchase a suite of fixtures and background material, plan on spending up to at least four times that amount. Many major bathroom rehabs, with a gutted interior, new fixture placement, and new windows, can cost up to eight times as much as a minor makeover. The most expensive upgrades are those that include premium fixtures or architectural changes to your home's floor plan. Keep in mind that prices vary by region and fluctuate with the marketplace, so these are only ballpark figures.

Furniture inspired vanities can make a cottage style statement. Here, the white bureau-like vanity bridges the visual gap between a restored clawfoot tub and a modern glass shower enclosure.

BELOW Beadboard is a classic element of cottage style. Sold by the plank or in paneling, this affordable material adds architectural charm to blank spaces.

ABOVE LEFT
To keep remodeling costs down, try to keep fixtures in the same location. Added-on beadboard that was given a sunny coat of paint brought this cottage bath up to date.

ABOVE CENTER
Judicious use of costly materials lends a high-end look without the high price tag. Simple cuts of marble create sink and tub surrounds that give this bathroom a touch of elegance.

RIGHT The most pleasurable bathrooms aren't the most expensive ones. Little extras such as this towel warmer and bath-side table make visitors feel as if they're in a romantic suite.

budgeting for a bathroom

The good news is that renovating a bathroom is considered one of the highest-yielding investments you can make in your home, whether you plan to stay for a long time or resell in the near future. Your bathroom is a space you frequent daily, so designing one that pleases you and meets your family's needs comes with a steady payoff. And should you choose to move on, prospective buyers are willing to spend more for homes that have modern, well-designed bathrooms.

TOP LEFT Furnishing your bathroom with vintage finds pays off in style and savings. This refinished clawfoot tub and funky Indonesian mirror make creative companions. The shelf unit is actually the top of a cupboard, mounted upside down.

ABOVE A compact master bathroom can save you money simply because fewer materials are involved. In this suburban townhome, the owners chose granite and maple because only a minimal amount was required to fill the space.

ABOVE Rethinking the use of a piece of furniture is one way to ease the budget. An unfinished sideboard looks at home in this small country-style bath when fitted with simple porcelain basins.

MONEY-SAVING TIPS

ave dollars in the end by planning ahead. Go through the design process first and choose everything you want, from fixtures to lighting. This will help you define your budget and prevent hasty (which can translate into costly) decisions in the thick of the process.

- Decrease the cost of your remodeling project through your choice of products. Do some legwork and shop for the best prices to determine if you can get the same look for less.

- Consider how labor intensive some design choices can be. For instance, installing tile versus cultured marble sheeting.

- Be creative and keep an open mind. In the design world, there are always multiple ways to solve a problem. Contractors are versed in discussing options.

- Stay inside your home's footprint. Adding square footage to your home increases your project's cost significantly. Consider borrowing space from a neighboring room or even a linen closet. You can grab back the lost storage by finding small niches between wall studs for shelving. If you really need to add space, consider a small bump-out of two to four feet to avoid changing the foundation or roofline of your home.

- Opt to reglaze a tub instead of replacing it, if it is in relatively good condition. This can save you more than half the cost of a tub replacement.

- Define what you truly want and need in your bathroom. Sometimes installing a double sink in an existing bath works just as well as adding a new bath.

Major remodeling projects call for an architect's expertise. One of the best ways to communicate with whoever you hire is by sharing photographs of rooms you love.

•finding a pro

Unless you are a seasoned and confident do-it-yourselfer, bathroom remodeling is best left to the professionals. Architects and general contractors will be familiar with the building codes and the process of obtaining a building permit, which vary from city to city and are generally required when structural work or the basic living areas of a home are changed. Depending on the size and complexity of your remodeling project, there are a few options for you to explore before finalizing your plans.

A General Contractor Modest home improvements don't require professional design services and can be handled by an experienced contractor. Seek out a home improvement contractor that has an established business in your area. Ask to see references from past clients in your community. Check with the government Consumer Affairs Office and the Better Business Bureau to ensure there are no complaints on record for the contractor.

Many states, but not all, require contractors to be licensed and/or bonded. Contact your state or local licensing agencies to ensure the contractor meets all requirements. Ask

the remodeling contractor for a current copy of his or her license. You should also ask to see a certification of insurance to learn the name of his or her insurance agency and verify coverage. Most states require a contractor to carry worker's compensation, property damage and personal liability insurance.

A Design/Build Contractor Design/build is a concept that benefits homeowners with their remodeling project by providing both quality design and construction services within the same company. This specialty contractor will be able to see your project through from start to finish, keeping design, engineering, and budget in mind. Many times, you can find these services through a home improvement center.

An Architect Major remodeling projects require construction drawings, which will help you to define your contract and local permit needs. If the professional you've chosen doesn't offer design services, you may be best served by finding a licensed architect. It's best to work with an architect experienced in remodeling, as he or she will be more sensitive to the often surprising challenges that pop up when you remodel within an existing structure.

ABOVE Converting existing bedrooms into bathrooms can be a simple endeavor that only requires a contractor with a background in plumbing. Simple side-by-side pedestal sinks and freestanding storage made this conversion less complex.

LEFT Before meeting with a professional, meet with your family to discuss your needs and wishes for a new bath. For this busy family, storage is a virtue.

defining your needs

●●● IN THE PLANNING PHASE, SPEND SOME TIME THINKING ABOUT WHO WILL BE USING THE BATHROOM and what the priorities are for the space. This can help you funnel your budget in the right direction. Will it be a room that is mostly for guests and will see little use? If so, design rather than durability may be your priority. Or, if this is the hub of your family's morning routine, layout and storage are the hot commodities. If you've been dreaming of a master suite retreat that waits to pamper you after a long day, you may want to check out the latest in steam showers and jetted tubs. If children are using the bathroom, there are special safety needs to consider.

Bathroom types break down most simply into five categories: master bath, family bath, children's bath, guest bathroom, and the powder room. The following pages will help you determine what features to include in your project.

An attic space, with its vaulted ceilings and low kneewalls, can mean wasted area. Making use of a dormer puts the awkward space to use by allowing a freestanding tub to be tucked under the eaves.

LEFT Creative solutions to common problems can result in unique bathroom style. For this family with small children, an oversized, apron sink (usually found in a farmhouse kitchen), and two pull-out steps means custom style and practicality.

ABOVE Fresh color and a lighthearted attitude can make any bathroom more appealing for children. A recovered dressing table stool gives the supervising parent a pretty place to perch.

LEFT Think about traffic patterns when you design a bath for two. By using a corner to their advantage, this couple is able to have ample counter and storage space in a small bathroom.

the master suite retreat

●●● PERHAPS IT'S THE LURE OF SEDUCTIVE NEW PRODUCTS or a natural antidote to higher stress levels among American consumers, but the master bathroom retreat is one of the most popular amenities—and rooms to remodel—in today's home. Really, who wouldn't want to have a private getaway that is only steps away?

There is a delicious list of amenities available for the modern master suite retreat. Some of the more popular features are massaging sprayers and faucets for the tub and shower, steam showers, saunas, soaking tubs, a separate niche for the toilet, audio and video systems, radiant heat floors, and towel warmers. These extras are tempting, but they need to be planned for. For instance, a large, jetted tub requires the proper floor support. Or you may need to add additional heating and electrical sources to fill the big tubs without draining the hot water tank.

Even compact master bathrooms can accommodate two people. This designer used the strategy of lining the walls with fixtures to keep the center of the space open. Etched glass shower doors and mirrors keep the space from appearing cramped.

LEFT When planning a master bath, consider how partial walls can create private areas for more than one occupant. This partition of glass block does the trick, without visually chopping up the space.

BELOW The right color palette will wash a master bath with serenity. Strong yet soft hardwood flooring plus limited accessories and pattern make a modern yet ordinary-sized bath both soothing and sleek.

gallery

the master bath

The essential elements of a high-style, high-function master bathroom include well-defined activity areas, space for two, sophisticated design, and pampering extras.

1 In this spare modern setting, ample space under the vanity and between the sinks, plus a no-threshold shower, means this bathroom can handle the busy morning routine.

2 When you create a calming master bath environment, you'll experience a serene start—and end—to your day. Though many of this room's surfaces are slick, its mood is warm and inviting thanks to honeyed hues and a textural view out the multiple picture windows.

3 Backing up the shower to a large, jetted tub is an efficient way to use floor space and share plumbing.

4 To make a small master bath appear larger, use smooth surfaces, minimal pattern, and plenty of glass. Featuring a large shower stall and spacious double vanity, this area doesn't appear crowded because of the well-chosen elements.

5 Prettiness has its place in a master bathroom that's adjacent to the master bedroom. Pull patterns, colors, even furniture styles and finishes into a bath space for visual flow from room to room.

4

5

plans for the master bath

The most popular layout for a shared master bathroom is one that allows two people to enjoy the space together, while providing privacy for individual matters. A separate area for the toilet and dual sinks on opposite ends of the room keep bathroom mates from bumping into each other during the morning rush hour.

Adjoining dressing rooms, walk-in closets, greenhouse bump-outs, or exercise rooms are all at the top of the most-wanted list of master bath amenities.

BELOW One way to carve out a water closet from a bath space is to create a small hallway. In this galley space, a shower was installed on the other side of the wall from the tub so that the plumbing wall could be shared. Further down the little hall is another niche for the toilet.

FACING PAGE TOP Create a room within a room to add privacy and break up a large open space. Columns add architectural oomph to this space, while supporting the glass walls of a large shower stall.

RIGHT Separate sinks are considered a must in today's master bathrooms. For a bit more elbow room—and storage—these homeowners added a slim vanity space between the sinks.

BELOW Keep the commode behind closed doors for privacy in a two-person master bath. In this light-filled space, clerestory windows and an arched wall create a focal point tub, while the humble commode hides behind a door at the left.

the family bath

● ● ● WOW, HAVE TIMES CHANGED! The average American house size has more than doubled over the past 50 years; it now stands at about 2,400 sq. ft. Back in the 1950s, it was considered normal for the family to share one full bathroom, which most often scooped up only a 5-ft. by 8-ft. space from the floor plan.

Are you faced with a room that seems trapped in time? There are many ways to bring it up-to-date in both design and function. If you're considering a compact family bath, put durability, ease of cleaning, and storage at the top of your must-haves list. As a rule of thumb, remember that matte surfaces show water spots less than shiny finishes. So you may, for example, want to select brushed chrome rather than the high sheen type. Also, acrylic surfaces are more mold and mildew resistant than porous surfaces, such as ceramic or porcelain.

RIGHT Consider your family's needs carefully. This family determined that storage and order is more important than a second basin. Custom cabinetry that stacks from floor to ceiling gets the job done, while creating a private vanity space.

BELOW Simple changes can "buy" space in a compact family bathroom. This narrow counter frees up floor space. The half in/half out basin also adds needed curve to all of the straight lines.

ABOVE Think outside of the box for more function in a family space. A curved built-in offers niches for towels and plumbing while emphasizing the curved lines also found in the sink and tub.

FACING PAGE Steeped in natural light and vintage style, a simple family bathroom can be charming and practical. This busy California family makes use of double pedestal sinks, a shower, and a tub in this high-functioning room.

• the compact family bath

When approaching the layout of a family bath, keep the concept of compartmentalization in mind. It's a big word but, simply put, it means that a few strategically placed walls can make one bath function as efficiently as two. Dividing a room into sections means more people can use the space with efficiency and privacy.

For example, think about how your shower and toilet might be walled off from the sink and vanity area. Another great way to keep everyone moving efficiently is to factor in ample storage. Separate but equal storage in a shared bath can make morning routines flow more smoothly.

RIGHT Smart space savers can leave enough room for family togetherness. A small vanity with a demilune sink looks great and offers storage where there's just a sliver of floor space.

FAR RIGHT You don't need walls of storage to keep order in a family bath. This large hanging basket can be used for extra towels or toys.

BOTTOM Necessity can be the mother of invention when it comes to squeezing in a second bath for the family. This family created a small bathroom in the corner of a child's room.

FACING PAGE Factor in extra sinks and storage to keep a family bath a peaceful space. Two étagères add structure and storage to this small, square space.

the children's bath

● ● ● THE BEST WAY TO DESIGN A CHILD'S BATH SPACE is to get down to their level, literally and figuratively. Try lowering your sightline to meet theirs and get a sense of how you can create a space that is tailor-made to little people's needs. If lowering the counter doesn't make sense, use a sturdy stool to give them a lift to the sink and mirror.

Emotionally, children are quite territorial. So for two kids sharing one room, you might consider how separate but equal personal spaces, such as cabinet and drawer space, coaxes kids to take more pride in themselves and their home. Letting children in on choosing colors and accessories is one way to give them a sense of ownership over the space.

Nonslip rugs, bright towels, and places to sit for dressing are useful features in a child's bath. Don't forget accessories that hint at fun, and in this case, the Cape Cod location of this cottage bath.

ABOVE Keep storage low to the ground to make it accessible to children. Happy Susan Sargent® fabric makes a pretty and practical skirt that hides baskets of bath essentials. It is simply fastened to the wall-mounted sink with hook-and-loop fastening tape.

SAFETY IN THE BATHROOM

W ithin every innocent bathroom lurks danger not found in the rest of the house. Keep these trouble spots in mind when planning a bathroom that a child will use:

- **Hot water.** Set your hot water tank temperature at no higher than 120°F and it will be warm enough to bathe in, but not hot enough to scald. Here's a quick test: You should be able to hold your hand under the water with only the hot water faucet turned on.

- **Electricity.** Cover all electrical outlets with safety plugs if there are children less than 10 years of age in the household. Update your system to include ground fault circuit interrupters (GFCI) on all outlets, light fixtures, and switches. Never leave anything plugged in (such as razors, curling irons, or hair dryers) when you are out of the room. No electrical outlet, fixtures, or switches should be within reach of a person in a tub or shower.

- **Slick surfaces.** Most bathroom blunders occur when people are climbing in and out of the tub or shower. Equip your bathing spot with grab bars, a bench, non-slip surfaces or mats, and low thresholds. All bath rugs should have a rubberized, slip-proof backing.

- **The medicine cabinet.** If a child is old enough to climb, he or she is old enough to explore the medicine cabinet. Keep all medicines and toxic cleaners behind locked doors.

- **Sharp edges.** Choose rounded corners on countertops and cabinetry. Rounded, oversized hooks are also advisable.

- **Standing water.** Toilets and filled tubs invite trouble. Keep a lock on the toilet lid and always close the bathroom door if you must leave a drawn tub unattended. Speaking of bathroom doors, make sure you have one that can be unlocked from both sides.

FAR LEFT Personalize a child's room to encourage pride of ownership and, just maybe, happier bath times. This sunny yellow space beckons with step stools that cleverly store bath toys.

LEFT Real footprints and freestyle flowers, along with the user's name and basic stats, stamp this bathroom with personality.

the guest bath

● ● ● IF YOU'RE LUCKY ENOUGH TO COUNT AN EXTRA BATHROOM IN YOUR HOME'S FLOOR PLAN, chances are it pulls only part-time duty. Whether the room sees only infrequent visitors, or is on a regular rotation of family members and friends, the goal of a guest bathroom design is to welcome and accommodate your overnight housemates.

When deciding on a new look for your guest bath, comfort should be the main goal. Warm lighting, a cosmetic color palette, and thoughtful extras such as hooks for hanging clothes or a basket of sundries go a long way to making a guest feel at home. Most homeowners don't feel the need to go all out on a second bathroom. Rather, selecting items that look great with a once-over and can become spit-spot on a moment's notice make the best choices.

ABOVE Keep a few drawers open in your guest bath to allow visitors to move in and feel at home. A separate bureau by this pedestal sink is handy for clothing and bath accessories.

RIGHT Don't forget to put some personality into a secondary bathroom. Once a dreary brown dining server, this small buffet is the perfectly sized vanity for a compact bathroom. Also given a second chance, old ceiling tins make clever wall tiles.

LEFT It's the thought that counts in a guest bathroom. In this Boston cottage, calming water-based hues and a stack of thirsty towels are equally pampering elements.

BELOW Little-used bathrooms allow you to experiment with materials that might not stand up to daily wear and tear. In this guest bath, teak counter keeps the look clean while offering the warmth of wood.

• guest bath extras

Think about the most memorable hotel rooms you've stayed in as you determine the look and level of luxury in your bath. For a short stay, having the most high-end materials is less important than the special touches: a lighted makeup mirror, cubbies for keeping toiletries neat and accessible, and a thoughtful stock of necessities like new toothbrushes, luscious lotions, or makeup removal pads. The comfort of a plush rug underfoot or a bathtub pillow and stack of current magazines also shows you care.

TOP A little romance is all you need to turn a simple bath into a memorable one. A bath tray of pretty oils and a fanciful towel—along with an artful wall of toile—are transporting elements.

RIGHT A few simple extras can let your guests know you care. A pretty cosmetic mirror, plenty of hand towels, and even an alarm clock add graciousness to this small bath.

more about...
AMENITIES FOR A GUEST BATH

1. Soft, plush towels. Evaluate your towels to make sure the ones you're sharing with your guests haven't seen better days. Purchasing a couple of new sets doesn't represent a big investment, but does represent thoughtfulness.

2. A cluster of candles, along with a stack of special books, encourages your guests to relax and bathe awhile.

3. A water pitcher and glasses will prevent your guests from stumbling down the hall to retrieve a nighttime thirst quencher.

4. While you're at it, a nightlight that shows the way to the bath will also keep guests from going bump in the night.

5. A small basket with travel size soaps, shampoos, hair and dental care items, and snack items will always be appreciated.

6. Extra hooks for hanging clothes or makeup bags will help your guests stay organized and keep their things close at hand.

TOP LEFT Happy colors and thoughtful accessories are welcome extras in a guest bath. These products are tailor-made for the male guest.

LEFT Tucking in a small dressing table turns an ordinary bath into a guest oasis. This painted vintage find, along with a medical office stool recovered in toweling, is a quaint place for guests to stash cosmetics.

gallery

the powder room

The obvious challenge in designing a powder room, or half-bath, is gaining privacy for visitors, plus determining a design that flows with the rest of your home—all in a tight area. To ease privacy concerns, plan for the powder room to open into a hallway or vestibule rather than directly to a public room. Layer some textiles into the space for sound absorption.

When choosing your paint, fixture, and surface colors consider the overall feeling you want for this little room. You may want the dazzle and drama of dark shades and gleaming surfaces. Or you may decide that visually opening the space with a clean, light palette goes best with the mood of your home.

What starts out as a small cube of space can take a flight of fancy with delicate patterns and touches of gilt.

To make a narrow space appear wider, resist the urge to use pint-sized pieces. This hefty vanity stretches wall-to-wall to counter the tunnel effect.

The basics can look beautiful when given a shot of strong color. While a full dose of rich blue would have engulfed this space, a half wall adds the right amount of color and character, while silhouetting the familiar form of a pedestal.

Vanities with the look of freestanding furniture are perfect for a powder room that is adjacent to living areas. With its marble top and traditional fluting, this vanity has classic appeal.

Don't be afraid to show some sass in a small space. A collection of vintage team portraits and a thoughtful layer of towels add hints of fun without spoiling the stateliness of this half bath.

planning your space

● ● ●

ONCE YOU'VE FIGURED OUT YOUR BATHROOM'S BASIC FOOTPRINT AND how it will be used (and by whom), it's time to think about the specific layout, or floor plan. The best way to get started is to do some planning on paper. If the placement of your fixtures is limited by plumbing concerns, consider those first, or discuss the cost of moving plumbing lines with your contractor. There are room arrangement specifications to consider as you begin to plot the placement of things. These tried-and-true decorating guidelines mean more efficient and comfortable movement through all the distinct activity areas within the space.

In this chapter, we will cover these room arrangement rules, as well as explore creative ways to work within different types of spaces. We'll also discuss ways to borrow or find existing space in your home to make way for your bathroom project.

Having a small bathroom doesn't mean you have to skimp on the luxuries. This 8-ft. by 8-ft. space holds a jetted tub and tiled shower, perfect for daily pampering.

•floor plan options

The folks at the National Kitchen and Bath Association provide consumers with information on planning a bathroom space as determined by both physical comfort and established building codes. Review these guidelines before you put pencil to graph paper to decide what types of elements your room can handle and how to organize them.

The first floor plan (below, left) illustrates that the minimum space of at least 30 in. should be provided from the front edge of each fixture to any opposite fixture or wall obstacle. Current guidelines for bathroom clearance needs are based on universal design concepts, meaning that the standards are developed with special needs individuals in mind, especially those in wheelchairs. This is one option for including proper clearance in front of a shower.

The second plan (below, right) shows standard clearance requirements for a bath with a tub with a built-in seat. The seat should be at a depth of at least 12 in.

COMPACT PLAN

A plan for a compact bathroom includes a 15-sq.-ft. shower and a door that swings out to save floor space.

WET AREA PLAN

A bathing area houses a 5-ft. tub as well as a large shower. A trench drain at the shower door keeps water contained. The shower measures 5 ft. by 6 ft.

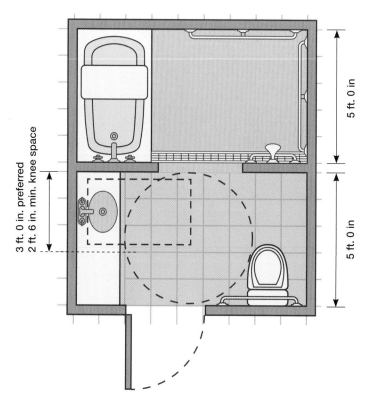

Though you need to keep in mind clearance standards for use of bathroom fixtures, you don't need to limit your design to standard floor plans and furnishings. In this basement bath, cabinetry built in around a freestanding tub makes the space unique and functional.

•planning for fixtures

There are certain minimum space requirements for placement of each fixture within your planned bathroom space. These standards are based on comfortable use of every area within the space. Code requirements must be considered as well and vary from location to location. Any contractor you choose should be well-versed in the local codes and should be able to provide written codes for you.

1. Door Clearance

Allow for at least a 32-in. doorway and for clearance when the door is open at 90 degrees.

2. Sink Placement

When positioning your sink, consider elbow room by measuring the center of the basin to a sidewall or tall obstacle. The distance should be at least 20 in. For a double sink, the distance between the centerlines of two basins should be at least 36 in.

3. Vanity Height

Vanity height is an important issue, and really depends on the comfort of the user. Most users feel comfortable with a height of between 32 in. and 43 in.

4. Shower Size

Plan for a shower to take up a minimum of 36 in. by 36 in. For optimum comfort, consider a larger space.

5. Toilet Placement

For comfortable use of the commode, the distance between the center line of the toilet and any wall, fixture or other obstacle should be at least 18 in.

6. Ceiling Height

The minimum ceiling height should be 80 in. when the ceiling is over a fixture. Shower heights should also be at least 80 in.

DOOR CLEARANCE

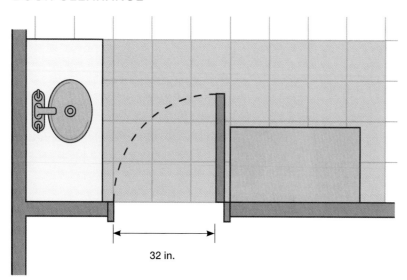

32 in.

SINK PLACEMENT

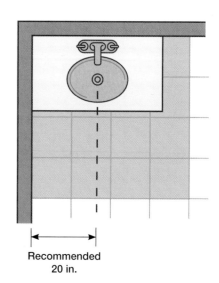

Recommended
20 in.

VANITY HEIGHT

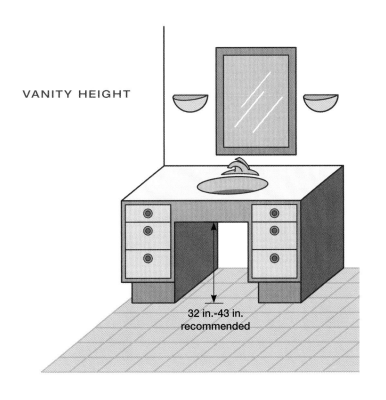

32 in.-43 in. recommended

SHOWER SIZE

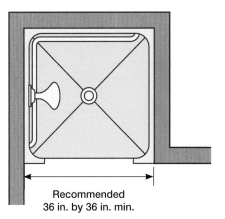

Recommended 36 in. by 36 in. min.

TOILET PLACEMENT

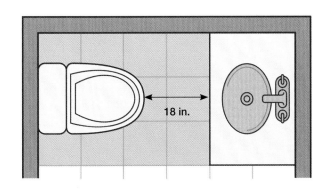

18 in.

CEILING HEIGHT

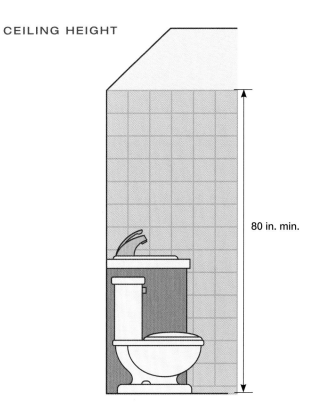

80 in. min.

universal design

● ● ● AS YOU PLAN AND PLAY ON PAPER, keep the present and future needs of all family members in mind. Chances are, you or a family member will undergo a physical transition at some point in your life. Whether it's a temporary impairment, like a broken bone, or a permanent disability, you or one of your loved ones will likely be faced with physical changes and challenges. The simple process of aging naturally increases our dependency on others. Universally designed features allow us to move through these changes and still enjoy equal access and independence.

You can make the room accessible to more users if you simply leave the area under the vanity clear for wheelchairs. This vanity still offers storage with drawers and a sliding mirror that reveals a medicine chest.

In European bathrooms, there is a concept called the wet room where the entire floor is fitted with drains and there is no separate shower floor. In this bath, it's clear to see how this idea fits naturally with universal design concepts.

Following the universal design guidelines doesn't mean your new room has to have all the style of a hospital bathroom. This modern space gets its high style from a wash of charcoal paint and black cement floors.

•the specifics of universal design

When building or remodeling, it's more cost effective to add many universal design features during the planning stage, when they can be built for little or no cost. Their addition at the start of your project precludes the need for future retrofits, which are always more costly.

RIGHT Fitting a shower with lever controls and multiple height or handheld sprays means your bathroom is ready for everyone. This shower also has a tile bench that looks great while providing comfort for users who want to get off of their feet.

BELOW Wide passageways and a floor absent of multiple levels and thresholds is key to universal access. This slate-covered bath also offers clear space under the vanity, and handheld shower attachments—all while keeping its cool, modern look.

PRINCIPLES OF UNIVERSAL DESIGN

by definition, universal design is "the creation of products and environments meant to be usable by all people, to the greatest extent possible, without the need for adaptation or specialization." This concept is not reserved for those with special needs.

The intent of universal design is to simplify life for everyone by making products and environments usable by as many people as possible, at little or no extra cost. Here are some key principles to keep in mind, as determined by the Center for Universal Design at North Carolina State University:

- Provide the same means of use for all users: identical whenever possible; equivalent when not.

- Accommodate right- or left-handed access and use.

- The design minimizes hazards and the adverse consequences of accidental or unintended actions.

- The design can be used efficiently and comfortably and with a minimum of fatigue. For instance, it allows the user to maintain a neutral body position.

- Appropriate size and space is provided for approach, reach, manipulation, and use regardless of user's body size, posture, or mobility.

- Provide a clear line of sight to important elements for any seated or standing user.

- Make reaching to all components comfortable for any seated or standing user.

- Provide adequate space for the use of assistive devices or personal assistance.

finding space for a bathroom

●●● MOST HOMES OFFER OPPORTUNITIES for adding or upgrading bath space without adding on. To see if your home offers such an opportunity, walk around with a creative eye. Obvious places to consider are undeveloped areas such as the basement, attic, or a porch that can be enclosed. Harder-to-recognize opportunities include hallways, closets, and empty corners of existing rooms.

Once you've found potential space, check more closely to see how practical it would be to convert it for use as a bath. If the new space backs up to an existing bath or kitchen and both can share a plumbing wall, you'll potentially save hundreds of dollars by not having to extend water, electrical supply lines, drains, and vents from another area. Of course, you'll probably need to run a new 20-amp circuit with GFCI protection. A whirlpool tub will also require its own electrical circuit. Wiring and plumbing will be easier and cheaper if the area beneath the space being considered for the new bath is a crawlspace or a basement with an unfinished ceiling.

ABOVE Putting a corner to use as a shower site means this homeowner found a niche for two ample vanities. Horizontal lines created by accent tiles draw the eye smoothly around the room so design doesn't appear choppy.

RIGHT If you are low on floor space but want a high-style bathroom, borrow principles of modern design. Rule number one: Less is more. A simple shower stall leaves room for a double glass vanity. A large mirror with inset lighting and a side-wall medicine chest keeps the interior open and light.

Finding space for a bathroom might mean borrowing closet area from a bedroom, as in the case of this small bathroom. By mounting the vanity above the floor, the room appears airier and, therefore, larger.

•found space considerations

You'll also want to look for venting possibilities (a code requirement in most locales): a window, exterior wall, or easy access to the roof. But if the space you've found doesn't have some of these budget-saving advantages, it isn't necessarily disqualified.

Another factor to consider is what you're giving up to have this room. Most found space involves a compromise: Gain a bath, lose a closet, for example. Since most houses don't have enough closets, you'll have to think long and hard about the tradeoff. Perhaps you can give up some bedroom space to replace the lost closet, or use an armoire for storage in a guest room. Consider all the options and consequences before you commit to the project.

Consider giving up part of a bedroom to make a charming bed and bath suite. These Iowa homeowners needed an extra bath in their 135-year-old farmhouse, so they carved a slice from an existing bedroom.

RIGHT If you get creative when installing bathroom plumbing, you might be surprised how many options open up to you. In this farmhouse bath, plumbing was installed under a raised floor when wall space wasn't available.

ABOVE Even the tiniest spot can function like a full bath with scaled-down fixtures. In this full basement bath, a mini wall-mounted sink leaves enough room for a shower to its right. A commode sits just in front of the shower wall.

MAKING A SMALL SPACE FEEL LARGER

So, you have big bath dreams but only a modest space in reality? There are ways to make a small bath look and live larger than life. But these tricks of the trade only go so far. First, determine if you have the minimum size requirements for a well-designed, well-functioning space, as outlined by the National Home Builder's Association: Powder room, 18 sq. ft.; bath with shower and no tub, 30 sq. ft.; bath with shower and tub, 35 sq. ft.

When faced with a pint-sized space, here are some design ideas for making it appear and function like a larger one:

• Stick with a light and airy palette. Walls and floors should be a light color, without busy pattern or too much contrast. Choose a simple shower door or shower screen. Stay away from overly ornate designs, as they fight visual openness.

• Keep vertical lines to a minimum and enhance horizontal lines. Consider tile borders and accents to strengthen your room's horizontal feeling.

• Use mirrors to make walls or structural barriers disappear into reflective surfaces. Mirrors also bounce light around the room to open up corners and niches.

• Repeat design elements—color, pattern, texture or scale—to make a place more unified and less choppy. For instance, in a tight space you might choose a single color for walls and fixtures.

• Shop for small but hardworking fixtures such as space-saving corner sinks, small toilets or tubs, pedestal sinks (if you have storage options), and small vanities or console tables (if you need the extra stashing power). There is more variety in the size of fixtures than you might think. For instance, old toilets are larger and less efficient than newer models. Even better, compact toilets have the same seat size as a round-front model and provide the comfort of an elongated seat.

Bathrooms that connect to a laundry or mudroom can be the handiest areas in the home. A touch of cobalt blue on this vessel sink links to accents in the mudroom and the stone floor ties the two hardworking spaces together.

TOP LEFT To add warmth and interest to a windowless space, think about cladding the area with grainy, natural woods. A basement bathroom conversion can take on a woodsy cabin feel. To make the small enclosure appear larger, the panels were installed horizontally.

TOP RIGHT There can be a surprising amount of space for a bath under the eaves of an little-used second-story room. Though kneewalls can present challenges, this owner solved any potential problems by punching out a display niche and using the nook for a toilet space.

BOTTOM LEFT When assessing your home for a new bathroom, think outside the box. This second-story sunroom has inherent charm with its raftered ceiling and bank of windows. At the end of the long narrow space, is a good-sized shower that used to be a closet.

BOTTOM RIGHT One trend today is an open plan master bed and bath suite. Sculptural, freestanding tubs can slip into extra space in an existing room. A series of artful windows adds light and enhances the modern design of the room.

attic baths

●●● CHARACTER-RICH ATTICS ARE READY ENVIRONMENTS in which to carve out a dreamy bath retreat. The biggest consideration is the plumbing—how to get the water up and down from this lofty space. Your best bet is to figure out where the plumbing is on the floor level under the attic and design your attic bathroom above, or very close to that. This will help to minimize costs. If things don't stack up for you or lack of wall space is a problem, consider raising the floor to house the plumbing. Just make sure to allow yourself enough headroom.

Also avoid the classic home improvement error: Make sure that you will be able to bring the bathtub, shower stalls, or any fixtures into your attic through the existing entrance. Imagine how disheartening it would be if you had the plumbing installed and then realized you couldn't fit the shower stall up the staircase to the attic.

Another challenge will be to figure out how to take the all slopes, angles, imposing beams, and peculiar details of an attic space and make them work for you. To avoid a choppy look, think about painting all of the walls and other nooks and crannies in a coat of creamy white or some other light shade for a soothing, monochromatic scheme.

RIGHT With their inherent charm, attic spaces make the most delightful and romantic bathrooms. Tucked behind bi-fold closet doors for privacy, this bathtub occupies one end of the attic, and the vanity and vessel sink are actually part of the bedroom area.

RIGHT A simple step up not only visually separates a bathroom from a sleeping area, it cleverly hides your plumbing. This older home didn't offer the opportunity to place the pipes in the walls, so it is hidden underneath the floorboards.

LEFT The most interesting attic baths contrast sleek modern style with the rustic rafters of the space, an idea that could work in basement bathrooms as well.

BELOW Since the attic is a separate and distinct space from the rest of the home, you can experiment with fun materials. Corrugated, galvanized tin creates an industrial chic ceiling in this modest attic bath.

special considerations for an attic bath

When considering an attic bathroom, you'll also need to think about light and ventilation. To let light flood in and steam escape, many homeowners install skylights to solve both issues in one delightful swoop. For sewer ventilation, you can go through the roof or, if you have roofing that you don't want to disturb for one reason or another, you can consider venting through a soffit.

If you have wood flooring you want to keep but are concerned about water damage, finish it with a water-repelling oil-based sealant or paint.

RIGHT Low storage along the kneewall also allows fixtures such as toilets or tubs to be placed where the ceiling is higher. A skylight and beadboard add more light and white to make the room appear larger.

BELOW The short kneewalls in a tent-shaped attic space might best be used as low-slung storage. In this Iowa bath, the cabinets include display surfaces as well as cupboards for stashing.

Many attics include windows at their peak, which make ideal focal points. The window is given more prominence here with a built-in window seat, under which is a great spot for storage.

room conversions

●●● IF YOUR HOME'S BEDROOM-TO-BATH RATIO is out of balance for your needs, consider borrowing space from a bedroom, or grabbing the whole room, to create an extra bathroom. Because you are staying within the home's footprint, this is a less costly option than adding on or bumping out to gain bathroom space.

Keep in mind some of the same money-saving strategies previously discussed, such as stacking or backing new plumbing against existing kitchen or bathroom plumbing, including water lines, drains, and vent stacks. You will also save dollars if your fixture placement limits plumbing to two walls.

This simple, country-fresh bathroom was once a bedroom. Putting the tub in the corner takes advantage of the sea views and breaks up the square space. A partial wall supports the wall-mounted sink and provides a private alcove for the toilet.

LEFT Freestanding tubs can be as artful as sculpture. With simple fittings, a place to bathe can be slipped into the most chic bedrooms.

BELOW Do you prefer a more traditional approach to romance? This slipper tub extends a pretty irresistible invitation from the corner of a guest room.

practical considerations

When converting an existing room to a new use, you will most likely need to obtain a building permit. Contractors typically handle this detail, but if you are doing the work yourself, you will need to investigate whether or not a permit for this type of remodel is required in your locale.

Also consider how the conversion will impact the resale potential of your home. You might check with a local realtor to see if a three-bedroom, two-bath home, for instance, has more buyer appeal than a four-bedroom, one-bath residence. As always, your decision should be based as much on your wants as it is your needs.

ABOVE Old homes are notoriously short on bathrooms. But many have small bedrooms that would make marvelous candidates for conversion, such as this Shaker-style bath which uses an old trestle table as a vanity.

BELOW An extra bedroom can become a luxurious master bath without moving walls. In this cottage, the homeowner kept the space open to create an amply-sized bath space.

When remodeling an entire room isn't practical, consider converting a corner into a place to wash up. This will ease the load on the existing bathroom—and possibly on the family's morning routine.

selecting
a style

● ● ●

PLANNING A NEW BATH AND DETERMINING YOUR BUDGET FEELS LIKE
due diligence, so deciding on a look you like is when the fun starts, right? Not
necessarily. For many homeowners, selecting an overall design scheme is a separate,
but equally daunting process. It includes picking out a palette, the style of the
fixtures, surface materials, and other design elements.

The design process is complicated by these many decisions and by the
substantial cost of a bathroom renovation. You're making investment decisions you
will, quite literally, have to live with for a long time.

For many, the first question to ask is the most difficult one: How do I know what
my family and I will like? As a first step, look around your home to determine which
elements should stay consistent from room to room, such as moldings, or a color
palette, or a style category. You don't have to match the rest of your home exactly,
but whatever style, color, or materials you use should complement the architecture
and furnishings found elsewhere in your home.

**Great style often is
a result of creativity
more than cash. For
intrigue in a boxy
space, a simple par-
tial wall lends a back-
drop to a slipper tub
while creating a sepa-
rate dressing area.**

Also, spend some time thinking about what looks you
gravitate to. Flip through the pages of this chapter to see many
of the current looks in bath design. Go through visual Web
sites or shelter magazines, marking what rooms appeal to you,
without thinking much about why you like them at first. Then
gather your images and consider what they have in common.

finding your style

● ● ● WHEN CREATING PERSONAL STYLE, the idea is to create rooms that not only please your eye, but satisfy your heart as well. So don't just consider how a room looks, but think also about the mood you want the room to conjure up. In other words, how do you want your home to feel? What kind of room do you want to greet you at the end of the day? Would it be a room that energizes and refreshes you? Or do you need a space that calms you? Do you crave more light? Or do you prefer a cozy corner to escape to? Think about these questions as you search for bath design elements.

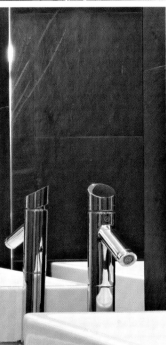

ABOVE Do you consider yourself a little bit country? With simple but chunky cabinetry and a casual linen window shade, this room might appeal to someone who likes an earthier life and style.

LEFT Or are you a little bit rock and roll? With a rug as boldly designed as a modern painting and walls painted to look like slate, this room is suited for those who like to keep it cool and calm.

WHAT STYLE ARE YOU?

One of the obvious first steps in creating personal style in your bathroom is figuring out just what your personal style is. Here are a few simple questions to ask yourself as you define your own look:

• What are your interests? If you like to go to flea markets, for instance, don't try to contain your style into a tailored and clean room. If you have a preference for modern art, you'll likely be drawn to a room with clean, contemporary lines. When you and your family travel, do you like to camp or stay in luxury hotels? Then consider how much nature, versus how much nurture, you want in this personal space.

• What is your fashion personality? Take a trip to your closet and take note of what styles you gravitate toward. Do your clothes tend to be buttoned up and traditional? Or do you prefer a more flowing style with all natural fabrics? Are you the romantic type that loves a bit of lace and frill? The answers to these questions are clues to your own design tendencies and sensibilities.

• What colors do you normally wear or decorate with? Do you choose bold patterns over basic solids? Do you dress in simple basics, then throw on an eye-catching piece of jewelry or a vibrant tie for a statement? Choosing a palette is the first step toward creating a pulled-together interior. Think about how color confident and pattern tolerant you are.

eclectic looks

● ● ● IT WASN'T SO LONG AGO THAT WE DECORATED BY THE BOOK—or by the era—slavishly following foolproof design formulas or period-perfect interiors. Now we want our homes to be an expression of our lives, our loves, ourselves. Creating personal style is the over-arching decorating trend today. And where better than the private rooms in your home to express your personality? But because this eclectic style is all about mixing rather than matching, it isn't necessarily the easiest trend to follow.

One secret to personal style success is to keep the decorating simple. Backgrounds that are kept pattern-free, for instance, will set off a mix of furnishings. A few bold accents make a stronger personal statement than a hodgepodge of smaller items.

A sense of nostalgia and surprise can stir up a sense of adventure in a room. Leather and lace team up in this California bathroom to create a romantic cowboy style.

TOP Why not put fine art in a bathroom? The idea today is to decorate to your heart's content. Here, a photograph is carefully framed in acrylic to avoid moisture damage.

ABOVE Juxtaposing two different styles, in this case country and modern, always creates an intriguing hybrid. Here, a primitive sideboard and a modern vessel sink and faucet make a strong style statement.

LEFT Industrial chic is a new style category that grew from the loft-living trend. A clean-lined steel table establishes that look in this small bathroom.

To douse a bath with personal style, bring in fanciful touches you might have reserved for other living areas in the home. In this Houston bath, artful lighting and a fabricated cement vanity make strong statements in an otherwise simple bathroom.

TOP LEFT Strong doses of color, when used sparingly, add sparks of personality to even pint-sized bathrooms. Slate tile and a modern faucet set a Zen-like scene in this California half-bath, which is interrupted beautifully by a grass green resin vessel sink and a zesty orange door.

ABOVE Restaurant supply stores are a great source for out-of-the-box bathroom materials. This commercial trough sink brings a fresh point of view to this bathroom, while picking up the sheen of iridescent glass tile.

ABOVE Creating an interior style you love means filling your spaces with objects that you love. A fanciful, antique gilt mirror makes a memorable statement in this small bath.

ABOVE Pull together a surprising mix of shapes and materials to design a bathroom like no other. In this Miami Beach home, a round tub throws this bathroom its first curve. Stone, tile, and glass block blend together beautifully, thanks to their soft colors.

classic and romantic

●●● DO YOU LIKE OLD MOVIES WITH HAPPY ENDINGS? Does nothing thrill you like a touch of ruffle and frill? Do you feel more comfortable wearing vintage clothing over Dolce & Gabbana®? Do you live in an older home with character-rich architecture . . . or wish you did?

You will likely lean toward design that is based in tradition but rich in romance. Some classic, romantic looks are marked by Victorian curves and a bit of excess. Other romantic designs are frilly, but lighter and airier in execution.

What romantic looks have in common is one foot in the past—the good old days in design when your rooms pampered you with little luxurious extras and plenty of details to entertain the eye. If a romantic look is alluring to you, consider ornate fixtures, traditional surfaces such as marble and porcelain, patterned wall coverings or fabric, and carved woods.

ABOVE To add a touch of romance to a bathroom, think pretty. In this clean cottage-style bath, the addition of a slipper chair dressed in a floral print fabric with a flouncy skirt is all it takes to amp up the romance factor.

RIGHT Traditional looks are at their best when they appear to have evolved over time. In this St. Louis home, subtle patterns and wood finishes that complement each other, but don't match to a tee, give the new space an aged look.

ABOVE A lush layering of accessories can signal a bathroom that has romance on its mind. The showstopper here is a heavily-carved gilt mirror, the perfect companion to a marble vanity on brass legs.

LEFT For a classic look, rely on classic materials. Gleaming marble and a simple, freestanding tub set the scene for a bathroom reminiscent of Roman baths of centuries gone by.

formal and refined

● ● ● NOT ALL TRADITIONAL LOOKS ARE FEMININE AND FRILLY. You can still look to the past for your design inspiration if you are a formal person who likes order and symmetry. Do you prefer antique furniture with straight lines and high polish? Are the clothes you wear tailored and trim . . . more Ralph Lauren™ than Tommy Bahama®?

A bathroom design that is orderly, balanced, and rooted in design eras past may be your cup of tea. You may want to stick with patterns crisp and symmetrical, and highly polished materials over matte finishes. Storage will be important to this neatly arranged decorating style.

ABOVE Don't hesitate to invite living room materials into the bathroom for a rich, traditional look. Here, wicker and rattan are pulled into service for a touch of colonial elegance.

LEFT Rich, polished woods are a hallmark of traditional style. This walnut vanity is a standout against subtle, creamy backdrops.

For a cleaner, more tailored traditional look, opt for minimal patterns. Substantial cabinetry and moldings mark this as the room of a traditionalist, but note how the raised panel doors and dentil moldings have a straight edge to them.

country styles

● ● ● COUNTRY STYLES HAVE LONG BEEN CONSIDERED AMERICA'S FAVORITE LOOK. Casual, natural, linked to our past, and inherently sentimental, country design reflects the way many Americans want to live. Like all other design statements, country schemes have been updated to reflect our more sophisticated and personal tastes today.

Are you a little bit country? Do you prefer jeans and cotton to fancy dresses and silk? Is your idea of a good time poking through flea markets for vintage treasure? For an evening with friends, would you rather take in a rodeo than an opera?

Today, there are as many expressions of country style as there are homeowners who love the look. The big umbrella of country takes in the fresh, airiness of a coastal cottage, to the clean Shaker style of the East, to the hand-hewn and woodsy mountain home. What the interiors all have in common are connections . . . connections to the past, to nature, and to the individual histories of the folks who reside within.

FACING PAGE LEFT The quirky niches found in country houses can be a bonus in bathroom design. Here, a second-story dormer becomes a light-filled stage for a freestanding tub.

FACING PAGE RIGHT The freshest country looks offer clearer color than the dark days of the past. The upper wall in this bathroom takes its color from a robin's-egg blue, country Windsor chair.

ABOVE Country design is as diverse as America itself. For a modern take, keep your backdrops clean and crisp. This gray and white backdrop (even the wood floor is painted) highlights the architecture of this Boston home.

•country style creativity

Creativity and resourcefulness are just two of country design's many charms. Using vintage objects in new ways creates style that's personal and fun. Best of all, decorating with found objects, period pieces, and humble elements can also provide a break for your design budget. When finishing out a bath country style, think about ways you can use salvaged pieces to your advantage. Old doors, lighting, beadboard, fabrics and even fixtures make intriguing and inexpensive features in your new bathroom space.

ABOVE Inventiveness is the mother of country design. This tub surround is cobbled from salvaged beadboard, while overhead the shower curtain slides on a track purchased at a hospital supply store.

RIGHT Don't limit yourself to drywall and tile in a new bath space. These clever homeowners carved out a private space steeped in style and fun by using salvaged barn boards and beadboard to cover the walls.

REFINISHING FIXTURES

eplacing a tub or shower are big-ticket items in a bath redo. The National Kitchen & Bath Association says an average bathtub removal and replacement will run you around $3,000. If your fixtures are worse for wear, but in good working order and in tune with your style—and the style of your home—consider refinishing them rather than removing them.

Porcelain, fiberglass, and cast-iron fixtures are all candidates for resurfacing, as long as the original finish isn't too far gone. If you see pitting and rough surfaces, you'll still see those bumpy surfaces after a new finish is applied. Generally, a professional tub resurfacing costs between $300 and $1,000, depending on where you live and whether the fixture can be restored in place. Some jobs require that the fixture be taken to a refinishing workshop. Sinks can also be refinished, but the cost of the job is roughly the same as buying a new one. So unless you love the look of the original, consider replacing it. Tile can also be resurfaced, but doing so will change the look of your tiled surface because the grout will also be glazed.

Old, cracked toilets usually aren't worth the trouble of a redo. Instead, invest in a high-efficiency toilet to reap water savings for years to come.

TOP LEFT Have a piece in storage that is currently out of favor? Don't rule it out; change its function and, possibly, its color. This rattan étagère was worse for wear. Then it got a shiny coat of black paint and new life as a vanity.

LEFT Creating great country style means thinking of old things in new ways. Once secured to a dressing table, this antique mirror has three times the charm, especially when sharing a focal point wall with a parade of collectible silhouettes.

clean and modern

● ● ● IF YOU THINK CLEAN AND MODERN MEANS COLD AND STERILE, you may want to take another look. Today's modern looks, when filtered through the trend of personal style, are simple and bold statements that aren't as studied, slick, and off-putting as in years past. The advantages of a spare and modern look are obvious; less ornamentation means less to clean—and clean around. Modern rooms are often subtle and understated, allowing the surroundings outside of the window to come into view. Given the inherently cleaner elements that go into the space, a new bath is a great place to shift to a more modern aesthetic, even if you tend to be more traditional in the rest of your home. Try cleaning up your design act a bit—you might like it.

So, are you a modernist at heart? Are you a person who feels more comfortable in an uncluttered environment? Do you love your up-to-the minute gadgets, and were you one of the first on your block to trade in your boxy TV for a slender flat-screen model? Do you wish you had more Calvin Klein® labels in your closet?

Today, modern looks are warmed by the use of earth-mined materials. This creamy limestone tile and countertop has an understated look; a border of pebbles adds a needed hit of texture.

LEFT Minimalist modern design only looks easy to accomplish. The secret is to make each object within the room a standout. Matching custom tub and basins are the stars here. After that, only a few textural extras are needed to complete the scene.

BELOW Symmetry is a principle of modern design. Note how that balance is carried through from mirrors to lighting to basins—right on down to the two matching stacks of towels.

LEFT Painting walls in innovative ways is a great way to heat up your modern scheme. Here, the different shades create a progression of color. The warm hues complement the black slate and marble, rather than add jarring contrast.

• what makes it modern?

Modern choices abound in today's market, and unlike in years past, sleek contemporary design is available for all budgets. And remember, many styles of furnishings can take on a clean, modern look if you simply dare to be sparer in your design execution. For a modern statement, keep your backgrounds clean and pattern free. Color is great, but the clutter found in too much pattern or too many objects will fight your serene scheme. Fixtures that are sculptural, but clean-lined, will fit right in with your contemporary look.

Is it modern, or is it country? Well, both. The vintage clawfoot tub has its roots in country design, but the minimal use of accessories brings it up-to-date.

TOP LEFT A small area can benefit from spare, modern design. Though a nondescript, boxy space, this bathroom is memorable thanks to the designer tub and intriguing accessories.

ABOVE Gleaming backgrounds make this bathroom feel contemporary. The glossy tile on the walls and floor bounce the light around from the adjacent sunroom, keeping the space light and airy throughout the day.

LEFT Architecture plays a big role in making a space a modern standout. This round window makes a stunning statement when framed within the small niche of an inset tub. The clean color enhances the bold statement.

ABOVE New circular tubs allow designers to put a fresh spin on bathroom design. Slate-colored stone becomes a sturdy base and vanity while above, a skylight and a series of simple bulbs echoes the design for a unified, modern style.

TOP RIGHT Tile is a versatile design material for creating one-of-a-kind baths. Here, tile isn't just used as a surround, but covers the entire custom frame to blend into the wall for a sleek look.

RIGHT Modern design relies on spare color schemes and spare floor plans. A classic navy and white pairing looks ultra-modern when unfettered by other colors or heavy pattern. The curved wall houses a shower and a simple, square basin.

LEFT Take a page from European bathroom design and create an efficient bathing space that includes the bath and shower and opens up directly into the rest of the room. This Swedish bath is at the ready for a simple rinse off or a long soak in the stone tub.

BELOW Inspired by the beauty of an egg, freestanding tubs in an oval shape are contemporary style at its purest. Large-scaled tile and a pared-down backdrop keep the look seamless.

downright rustic

●●● A NEW AWARENESS OF EARTHLY MATTERS HAS ushered in a fresh appreciation for natural materials and a more organic design aesthetic. Natural materials have inherent quality and a purity of design that appeals to those who are seeking a balance to the plastic and techno forces that are so much a part of our modern world.

Are you drawn to natural beauty? Would you prefer a walk in the woods to a stroll through a bustling shopping district? Would you rather wear flowing linen dresses to a belted rayon number, even with all those wrinkles to contend with?

Like its country design cousin, rustic design schemes are relaxed, natural, and linked to the great outdoors. The range of materials available in stores—stone, wood, cement, glass, and so on—is now wide and varied enough for you to choose and carve out your own personality within it.

ABOVE Rustic looks don't have to be dark and heavy. With its stucco walls, painted plank floors and stone basin, this attic bath is earthy enough, but the modern tub and limited accessories lighten the mood.

RIGHT Dark woods and natural stone create a foundation for a rough-hewn look. The ocean-blue tiles in this shower add a counterpoint to the heavier elements to freshen and brighten the space.

Exposed beams and latillas (small poles used to form a ceiling in traditional Southwestern design) top off an unpolished room. The chunky carved bureau converted to a vanity balances the ceiling and adds to the rusticity.

the green bath

● ● ● THE INTEREST IN ECOLOGY-CONSCIOUS DESIGN SHOWS NO SIGN OF WANING. As a result, manufacturers in the home industry now recognize that this is not a passing fad but a widespread design and cultural movement. From fixtures to surfaces, construction materials to cleaners, the array of green products waxes on, seemingly endless. The demand is also growing as consumers realize that being easy on the planet also translates into savings for the household budget. When you consider both the personal and the environmental benefits, going green is an appealing route.

ABOVE What makes a space eco-friendly? The use of natural materials rather than synthetics and plastics does, for one thing. Natural slate in this sun-filled bath space looks great and offers geothermal benefits.

LEFT Organic bath textiles add a final layer to a green bath. This bath also features a natural slate wall and streamers made from gathered shells.

TOP Green design uses woods harvested locally to avoid transporting. Famed designer Russel Wright built this home in Manitoga, New York according to green concepts of the day.

ABOVE Think about the cost of filling a large tub before you buy big. This small, circular tub also has a handheld faucet so you can bathe while the tub fills and use less water.

GREENER BATHROOMS

going green in your bathroom may be easier than you think. Here are ten things to think about as you plan your new bathroom:

1. Consider natural choices for flooring, such as natural linoleum (made from a blend of resins, oils, chalk, and cork), cork, and bamboo.

2. Try to find materials that are produced locally, or select renewable materials such as bamboo, wheat straw, or cork. Avoid wood products that contain formaldehyde to safeguard indoor air quality.

3. Shop for recycled materials, such as glass, rubber, and even wallpaper.

4. Select low or no volatile organic compounds (VOC) paints for better indoor air quality.

5. Avoid incandescent bulbs, opting instead for compact fluorescents (CFL) or LED bulbs. These new bulbs last longer and use less energy.

6. Install as many energy-saving features as you can, including low-flow showerheads, faucet aerators, and high-efficiency toilets. Look into the discreet urinals that are now on the market.

7. Use durable, sustainable materials to avoid replacement costs. Concrete and slate are two of the best bets because of their strength and versatility.

8. Smart window placement and coverings can save you energy and money. One way to make existing windows more energy-efficient is to dress them in layers, more layers in the winter to retain heat and the sun-control layer in the summer.

9. Consider an on-demand water heater. Also, gas water heaters save more water than electric. If your water heater is warm to the touch, cover it with an insulator.

10. Clean your bathroom with natural cleaners like vinegar, salt, and baking soda—or shop for the eco-friendly cleaners now widely available. These products aren't harmful to the environment when they go down the drain.

•green considerations

So, while you plan your bathroom makeover, ask yourself, is this the right time to go green? Consider this: The bathroom is one of the most resource-intensive rooms in the house. As the largest user of water and a major energy consumer, it's a great place to reduce your home's environmental impact. Manufacturers estimate that green products are on average 10 percent higher than the cost of non-eco materials, though the gap is diminishing as the popularity of green grows. Given the water savings you will yield, and the peace of mind that comes with using nontoxic materials in your family home, can you afford not to look into your green options?

For a pleasant way to be environmentally friendly, install skylights for ventilation, natural light, and solar gain. The owner of this master bathroom also used locally harvested pine and low VOC stains to create their focal point, the double vanity.

TOP Reclaimed woods and other natural materials are good green options. Note the copper top on this stone pedestal vanity. It will last forever with minimal care and will only get more graceful as it ages and oxidizes.

ABOVE Natural slate, concrete, and bamboo are easy on the environment, and on the eyes. That mix of textural materials comes into play in this modern bath, along with thermal pane windows that have been etched in a band for privacy but still let the warm sun beam in.

more about...
GREEN SAVINGS

t here is quite a bit of information about the economic cost of going green. But consider the other side of the coin: How much will you save? Here are tips from the Green Home Guide published by the U.S. Green Building Council.

Reducing water use is the most important thing you can do to make your bathroom environmentally sound. If your bathroom was built or remodeled before 1994, the toilet is using at least twice as much water as a modern model and your shower is using as much as triple today's standard. That matters because every gallon of water used at home consumes fossil fuel during its extraction, uses chemicals during purification, and emits chemicals into our air and water during the wastewater treatment process. Wasting water also reduces supplies available for fish, wildlife, and ecosystems.

Toilets. Toilets are the single largest user of water in the home, accounting for up to 28 percent of water use. Replacing an old toilet with a new model will instantly save water and money, up to $50 a year. A new 1.6 gallon-per-flush toilet costs roughly $300 installed, paying for itself in six years and cutting water use in half—a savings of over 180,000 gallons of water during that period.

Showers. Showers use 16 to 20 percent of an average U.S. home's water. You can replace a showerhead inexpensively and quickly, cutting water use up to 70 percent. While the federal standard dictates a maximum flow rate of 2.5 gallons per minute (gpm), 1.5 to 2 gpm works as well and saves more water.

Faucets. An average household's faucets account for 15 to 18 percent of indoor water use, and bathroom faucets tend to be the most heavily used. If you are installing new faucets, look for 0.5 to 1 gpm models. If you are retaining older faucets, install aerator heads.

exploring the art of bathing

● ● ● BECAUSE WE'RE PLUGGED INTO THE WORLD BY IMAGES AND INFORMATION ON TELEVISION AND THE INTERNET, there is rising interest in customs, rituals, and design found in other cultures and countries. In the bathroom, an intriguing ritual to explore is borrowed from the Japanese custom of bathing. It seems the Japanese have known for years what we are just discovering: The bathroom can be more than a room that refreshes the body; it can be a place to restore the spirit as well. Is it any wonder why we stressed-out Americans are fascinated by the art of the Japanese bath? The best part is, it may not take much to add a little Zen to your space.

The sensory aspect of a bathing space is not overlooked in Japanese-inspired rooms. Polished wood and stones bring the user closer to nature in this woodsy bath. The polished cinder block tub, accessed by a natural stone step, goes with the Zen flow.

more about...
TRADITIONAL JAPANESE BATHS

i n Japan, any time of day is the right time to take a bath—as long as you have enough time to devote to the meditative and relaxing experience. The actual tub in a Japanese bathhouse or private bath is much deeper than Western styles. It's designed so you can sit comfortably, up to your neck, in the hot water. Traditionally, there is also a sauna for drying after the bath.

Washing takes place in the bathroom before you get into to the hot tub. Typically, you sit on a low stool in front of taps and showers set into the wall of the washing area. The bathroom is warm and steamy so washing is pleasurable. The body is scrubbed with soap and warm water using a bowl and hand shower. Because you go in clean, the hot water is often used by more than one person and often at the same time for social and conservation reasons.

The bath itself is a sensory experience as the water is often infused with flower petals or aromatic touches. Remember that the point is to cleanse not only your body, but your spirit.

TOP LEFT
A shower can also use Japanese pre-cepts. This one is ready for a sponge bath with its low faucet and basin. Or visitors can opt for a rain-like shower from the rain-bonnet showerhead.

FAR LEFT Western design borrows from Japanese custom, but gives it a modern twist. In this sunken, slate tub, the bather can sit and wash with a low-flow hand sprayer while filling the tub for a soak afterward.

LEFT Can't find the right tub to fit your natural style? Make one. With wood as polished as a yacht's, this extra deep tub is a delightful ex-perience. True to Japanese custom, the bather first washes so the tub water can later be shared.

accessorizing the bath

● ● ● ONCE YOU'VE DETERMINED THE STYLE you like and have selected your fixtures and back-ground materials, don't stop decorating your new bath. The final layer of trimmings and extras is where you can add visual interest and express yourself.

Collections, mementoes of favorite trips or periods of your life, treasured gifts from special people in your life, photographs—all of these items give your home more character because of the meaning and memories that are attached to them. Other types of objects have inherent significance, such as a stunning seashell, a gathering of natural sponges, or flowers. Art, by its subjective and creative essence, is an organic way to express your taste, and give your eyes a treat.

A virtual flood of great products for the bath makes accessorizing fun and easy. Here, a low bench keeps good-looking grooming products close at hand.

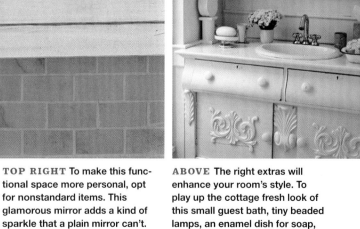

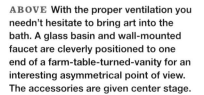

ABOVE With the proper ventilation you needn't hesitate to bring art into the bath. A glass basin and wall-mounted faucet are cleverly positioned to one end of a farm-table-turned-vanity for an interesting asymmetrical point of view. The accessories are given center stage.

TOP RIGHT To make this functional space more personal, opt for nonstandard items. This glamorous mirror adds a kind of sparkle that a plain mirror can't. Underneath it, a built-in shelf shows off more pretty pieces.

ABOVE The right extras will enhance your room's style. To play up the cottage fresh look of this small guest bath, tiny beaded lamps, an enamel dish for soap, and petite pitchers of flowers are called into service.

accessorizing tips

As in every other room of the house, the most important rule of accessorizing the bathroom is to know when to quit. Yes, adding objects of interest will create a beautiful environment. Just don't crowd the space or make your objects fight for attention. Leave room for you, and your eyes, to rest.

The best way to create an orderly display is to start with a large object first, then fill in around this focal point piece, staggering heights as you go. Play with your arrangement until it pleases your eye. And remember: One bold, beautiful piece always beats several mediocre ones. Keep studying published rooms and practicing placement with your own favorite pieces and you'll be a master accessorizer in no time.

ABOVE Select accent pieces that have something in common to avoid a cluttered look. By using only black and white family photos and green bottle glass, this farmhouse bath retains its calming personality.

RIGHT The best bath accessories are those that look pretty and perform a necessary function. A wire tray can hold lotions, potions, and sponges—and you might leave room for a good, long book.

TIPS FOR CREATING ARTFUL AND PERSONAL DISPLAYS

GET IT TOGETHER.
Don't spread collections around the room. You'll create more impact if they're grouped together to create one visual unit versus several little pieces. If you love small objects, for example, the best way to get them noticed is to corral them on a tray or mass them on a single shelf.

BE CONSISTENT.
When decorating with a variety of framed pieces, for instance, consistent framing will tie the individual pieces together beautifully. Painting dissimilar objects a common color is another way to achieve a pulled-together look. Again the goal is to gather accessories into groupings that can be considered a single unit.

LEARN THE TERMS.
Familiarize yourself with basic design principles such as balance, scale, repetition, and harmony. Look up the terms. They're pretty straightforward. The idea is to pull together a display that leads your eye through it in an orderly and pleasing way, with nothing jarringly out of place.

GET IN TOUCH WITH NATURE.
Flowers, ferns, shells, and stones are all objects with inherent beauty because they come from nature. Relying on touches of nature will softly fill in the blank spaces in your rooms and displays.

DESIGNATE A "PROP CLOSET."
Sometimes the way to improve your room is to un-decorate it. Get a fresh start by clearing the decks of all objects. Only put back what you truly love and store the rest. You might be amazed what a good thinning out will do for your space.

Parade your personal passions to make a bath space a one-of-a-kind experience. This history buff and antiques collector showcases some of his favorite curios in a small guest bath. The repetition of black elements keeps the room tied together.

tubs, toilets, showers, and sinks

● ● ●

OVER THE LAST COUPLE OF YEARS, THE ADVANCE IN DESIGN OF BATH fixtures in form and function has been nothing short of revolutionary. This revolution has been fed by design-savvy consumers like you, who are increasingly more conscious of what a few spa-like extras can do to improve your attitude, your health, and your day.

How would you like a bathtub designed to lower your heart and breathing rate through a combination of chromatherapy (color therapy), sound vibration, and air jets? Modern multi-faucet showers look like the human equivalent of a high-end car wash. They clean you, certainly, while they loosen up your body's knots and invigorate you at the same time. Today's sinks pop out of the box flaunting as many shapes, materials, and vanity options as you can imagine. Toilets with anti-bacterial surfaces and shapely silhouettes clean up after themselves—and look good doing it—while using as little as one-fourth of the water as their earlier counterparts.

Once again, the choice is yours. Let your budget, your needs, and your desire for a bit of personal pampering guide you through the following roundup of options for tricking out what will soon be your favorite room in the house.

Low slung, easy-to-enter tubs were first designed in Italy and have now become part of the high-end product mix for American buyers. This sleek beauty doesn't get in the way of a priceless view of San Francisco Bay.

bathtub beauties

●●● IN 1883, A WISCONSIN FARM-IMPLEMENT MAKER NAMED JOHN KOHLER began coating his cast-iron horse troughs with white enamel in search of a new type of product to sell. When he put four ornamental legs on it, the first enamel bathtub was born for residential use in the United States.

That simple clawfoot design was already evident in England, where royalty had enjoyed its use for centuries. It remains a popular classic, but the bathtub has traveled a long way, its evolution fueled by technological advances in materials and function, and by consumer tastes. Today's bathing spots can be as advanced, or as utterly simple, as you desire.

ABOVE Tubs designed to be sunken into a surround give a homeowner the opportunity to link their bathing spot with the architectural style of the rest of the home. In this California cottage, beadboard, creamy white tile, and simple cabinetry create a fresh bathing spot.

ABOVE The slipper tub got its moniker from its shoe-like shape. A higher back on one end allows its users to sit at a comfy, reclining angle. Whether antique or newly minted, these tubs seem to signify personal luxury.

cast iron and china plumbing fixtures were the only options 50 years ago, but times have changed and so have finishing materials. Still, you are basically faced with the decision to choose cast iron or metal-based finishes or synthetics such as acrylic and fiberglass. Here are some pluses and minuses on these options:

1. Today, traditional cast iron tubs are manufactured in virtually the same way they were a century ago. Molten iron is poured into a mold and, when cool, a thick layer of porcelain enamel is fused to the cast in a high-temperature oven. **The upside:** The enamel coating is hard and brilliant. The smooth-as-glass surface cleans well, but needs to be cleaned regularly to prevent buildup of dirt and soap residue. **The downside:** The enamel finish can chip and crack, which is best remedied by re-enameling the unit. The weight of these units can put your plumber in the chiropractor's office. However, you can shop for lighter metal-based units that have the same qualities of cast iron.

2. Acrylic and fiberglass have given manufacturers the ability to design and mold bathing units to their hearts' content. **The upside:** The lightweight units come in all sizes, shapes, colors, and shower/tub combinations. No grout used means low maintenance and mold-free. **The downside:** Without extra structural support (added by your contractor at more cost to you), these units can flex and possibly crack. It is difficult to make repairs to these finishes, though some scratches can be buffed out. Colors have also been known to fade over time, especially if harsh cleaners are used. Acrylic is more expensive and considered to be of higher quality than gel-coated units.

3. Gel-coated units are made by spraying pigmented resin onto layers of fiberglass. **The upside:** If the gel coating is thick enough (and you can determine its thickness by reading the maker's specs), this type of synthetic unit can offer a stronger finish than acrylic. **The downside:** If not manufactured up to high standards, these units are not as durable and fade-resistant as the other options.

4. Solid surface tubs are made of synthetic materials such as Corian® or natural stone such as marble. **The upside:** Solid surface materials can be molded or carved into almost any shape, leading to tub designs that have sculptural appeal. Synthetic materials are much lighter than natural stone. **The downside:** These sculptural beauties are among the most expensive.

TOP Enclosed tubs have the advantage of a shelf-like surround that conveniently holds inspirational items and practical ones. In this Portland area bathroom, the lively mosaic of wall tiles is balanced by the simple, singular touch of natural coral.

BOTTOM A variation of the clawfoot tub, this pedestal tub makes a focal point statement in a spa-inspired bathroom. The room is calming with its subtle water palette, its symmetry, and its glimmering surfaces of marble and glass.

Freestanding tubs have exposed pipes, and so are easier to install than built-in models. This enameled cast-iron tub can sit directly on any floor surface, which makes it a flexible design element.

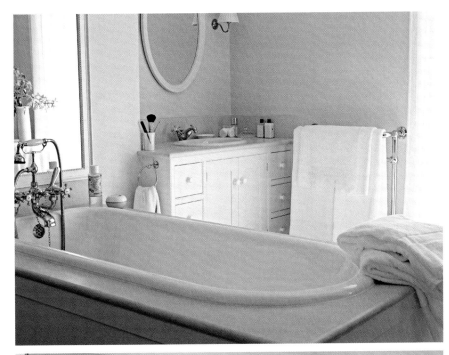

LEFT A built-in tub unit requires additional carpentry to enclose it, but the units themselves can be less expensive because they are designed with less finished surface than freestanding types.

BELOW If you decide on a freestanding tub you'll need to consider where and how to house the fixtures. This clever Iowa homeowner built a box for his low-slung resin tub so that the fixtures could be tucked away neatly without being mounted to a wall.

ABOVE Freestanding tubs come in all shapes, sizes, and materials today. Whether made of metal, such as this modern beauty, or other organic materials such as stone or wood, these works of art can be the most expensive option.

ABOVE Tile is a classically beautiful way to finish off a built-in tub. Decorative bullnose tile (or round-edge tile) finishes off the white, subway-tile installation with a flourish.

• bathtub choices

Hydrotherapy. Aromatherapy. Chromatherapy. You'd think you'd need an advanced degree to buy a tub these days. With all the bells and whistles, finding a tub with the right size, depth, and function for you and your family might mean trying a few on for size. Keeping within the parameters of your room, select different models that might work for you. Then get right into the tub on the showroom floor to determine what depth, size, number, and placement of jets are right for you. Settle back and adjust yourself to other comfort features such as armrests and head pillows.

The many options are yours to consider. One word of caution: Don't get a bigger tub than you need, as even the largest hot-water systems might run out of warmth before a super-sized tub is filled. (Not to mention the amount of water they consume.) No matter if your tub holds 50 gallons or 200, however, consider adding an inline heater to maintain a constant temperature while you're in the tub. You will also need to check the manufacturer's specifications to determine whether you need to reinforce floor joists to handle the tub's weight.

WHIRLPOOLS VS. AIR BATHS

now you have a choice between swirling waters generated by water jets or air bursts. Within those two options are several more to consider. Here are a few pointers to help you decide:

Water experience. Whirlpools offer targeted body massage with adjustable flow and direction. Air baths offer a full-body vibration experience, with multiple air ducts powered by a blower that warms the air to keep the bath a comfortable temperature.

Cleaning. Look for jetted tubs that offer self-purging or flushing functions to keep the jets operating efficiently. Many jetted tubs offer jets that are even with the tub wall, adding a level of comfort to the experience.

Bath additives. Many traditional whirlpool tubs eschew the use of bath oils, salts, and gels because their residues can clog the jets over time. The new air baths are accepting of most bath additives.

Working parts. Whirlpools rely on a pump that recirculates a mixture of air and water. Air baths inject warm air through a blower system.

ABOVE When you want luxury in a small bath space, consider enclosed tubs that can be placed at an angle. This smaller tub leaves room for a window seat that lets users enjoy the steamy warmth of the bathroom for a longer period of time.

RIGHT Promising "perfectly smooth air jets" and a blower with a built-in heater, this premium acrylic air bath from American Standard® puts comfort first. One advantage of an air bath over a jetted tub is its ability to regulate water temperature for those who like longer bath times.

FACING PAGE Kohler's® RiverBath® simulates real river sensations such as a waterfall, rolling currents and rapids. You get to select your sensations with the push of a button in a bath designed to hold up to three of your closest friends.

• custom baths

Today there are custom and ready-to-purchase baths that are chiseled out of stone or formed from stainless steel. They can be designed to hold deep wells of water or inset into niches enclosed by wood, stone, or windows. The bath has come a long way from its Roman roots—or has it?

Many of the extravagant marble and metal trough-style baths and sinks are legacies from the earliest days of bathing, when the Romans engineered and enjoyed extravagant bathhouses. The ancients gave us aqueducts and furnace-heated water, and the understanding that baths were not just for hygiene, but for overall good health and healing. Perhaps a luxury bath is just what the doctor ordered for you.

BELOW A light well in the ceiling and a water well in the floor make for a Zen bathing experience. This bathroom is inspired by traditional Japanese baths, which are deep enough to let a seated person soak up to his neck.

ABOVE Stone holds water and heat, making it a natural choice in bathroom surfacing materials. With the importance of eco-friendly design today, natural stone is becoming a popular choice.

TOP RIGHT A rectangular trough of granite becomes a deep well of relaxation in this solar-heated bathroom. The granite tub mimics the trough sink, also carved from the tough stone.

ABOVE European-style freestanding baths are making their way to our shores and being designed into modern bathrooms loved for their calming minimalism. This acrylic tub sits next to a barely-there glass shower.

RIGHT With a naturally-appealing egg shape, this Italian design takes the tub to its bare essence. Minimalist faucets and fittings that rise right up out of the floor don't encumber the spare design.

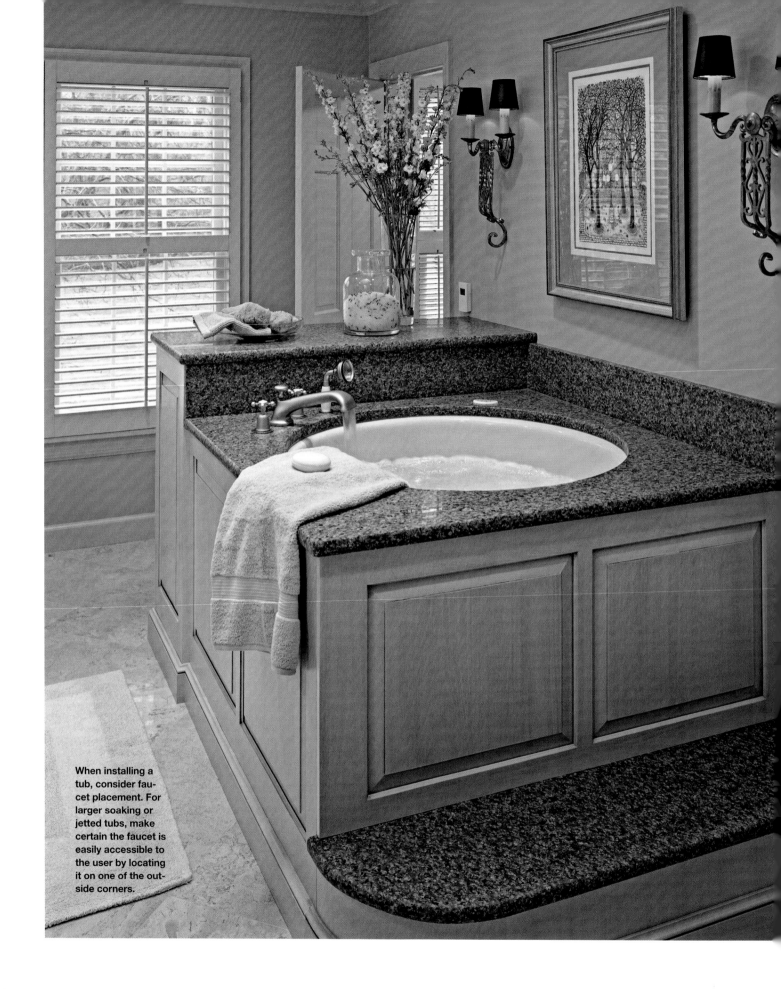

When installing a tub, consider faucet placement. For larger soaking or jetted tubs, make certain the faucet is easily accessible to the user by locating it on one of the outside corners.

FAR LEFT Wall-mounted faucets work well for custom bathtub installations or for tubs that aren't predrilled. The advantages of a simple wall-mount include flexibility in the height of the faucet and a cleaner tub ledge.

LEFT Gooseneck faucets were once reserved for the kitchen, but now bring their practical elegance to bathroom sinks and tubs. This brushed chrome model reaches well out into the tub basin to alleviate splashing and flooding along the tub's edge.

ABOVE Faucets in all shapes and configurations top off your tub design in both pretty and practical ways. This waterfall type keeps a low profile, while filling the tub at a faster rate.

LEFT New, stripped-down faucet styles are available for the modern freestanding tub. With fittings this clean and sculptural, who needs to hide them?

• specialty baths

Find the right bathtub to suit your style, comfort needs (today and 10 years down the road), and budget. Custom baths and jetted hydrotherapy tubs can easily become more expensive than a living room full of new furniture. But like a favorite sofa, a special tub can improve your daily life, physically and emotionally. Consider that the health benefits of water therapy have been known for centuries. The Greeks and Romans discovered the healing power of hydrotherapy, and we still believe in the benefits of water therapy on aching muscles and stressed out minds.

Today, you can select a tub to soak or to exercise in. Many models come with lower sides to make entry easier for young folks and seniors. You can choose from a wide variety of colors, shapes, and materials. The key is selecting the model that not only looks right, but feels right for you and your family.

TOP LEFT Clawfoot tubs come in lengths between 4.5 ft. to 6 ft., so be sure to select a comfortable fit. With its rolled edge, this classic tub is designed for reclining.

LEFT Stainless-steel tubs are considered the classic, Rolls-Royce of tubs—and they have the price tags to prove it! However, the stainless that clads this enamel slipper tub is a sustainable product that also keeps water warmer longer.

FACING PAGE Create your own hydro-therapy spa with a jetted tub you can design around. Here, a background of glass tile and slate—complete with a small waterfall in the pebbled niche—make this soaking spot good for body and spirit.

•showers with splash

If you have the space to spare, it's a good idea to design your bathroom with distinct showering and bathing areas. Compared to combination tub and shower units, a separate shower is generally larger and easier to enter and exit, making it safer and more enjoyable to use. Spaces that are used for only showering are easier to keep clean—no standing water or bathtub rings—and can be outfitted with benches for seating, multiple showerheads, and body jets or steam functions. There is a range of enclosure options for the shower area as well, including etched or clear glass, tile, or no enclosure at all.

Commonly known as the rain-bonnet shower-head, this large fixture produces the calming feel of a warm rain. Today, this classic has been made more water-efficient.

From hand showers to massaging shower-heads to ceiling-installed rain-bonnet fixtures, the idea today is to design your own mix, based on your own needs for physical and emotional comfort.

ABOVE The well-designed bath includes a shower that is an integral part of the room's scheme. Here, the typically squared-off shower unit gets some curve, making a shower curtain unnecessary.

LEFT A late-breaking innovation in showering is the "shower tower." In Kohler's version of this hydrotherapy system, a basin built into the shower floor collects and recycles water. Users adjust the intensity and direction of the jets.

BELOW Double showers are the ultimate in a fun and efficient bathroom. Using classic white subway tile and other vintage designs, this shower fits snugly into the corner and offers a showerhead perfect for every member of the family.

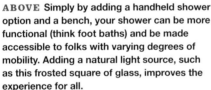

ABOVE Simply by adding a handheld shower option and a bench, your shower can be more functional (think foot baths) and be made accessible to folks with varying degrees of mobility. Adding a natural light source, such as this frosted square of glass, improves the experience for all.

ABOVE Modular shower units are the most cost-effective choice in shower options. This simple corner unit has been integrated seamlessly into a white-paneled wall design to achieve the look of a custom-built shower.

•shower surrounds

There are many options for finishing off your dream bath, but your decision-making should start behind the scenes with water-resistant wall material. The most common type of sub-wall is water-resistant drywall, sometimes called "greenboard" because of its color. Cement board is another option; it will add more durability, water resistance—and expense—to your project.

The surface material itself must be waterproof (preventing water penetration completely), not just water-resistant (which resists but doesn't prevent water seepage).

Various manufacturers offer prefabricated surrounds made of fiberglass, acrylic, vinyl, plastic laminate, or synthetic stone that are assembled on site. A readymade option is a multi-piece shower surround kit, which can be assembled inside the bathroom and is available in a variety of styles and colors. Though you can buy a one-piece surround (typically molded of fiberglass, gel-coat, or acrylic), you'll need to measure carefully to make sure the product can get through doors, down hallways, and around tight corners.

CREATING YOUR OWN SHOWER SURROUND

if you decide to have a custom bath or tub surround, you have a choice of surface materials.

Solid-surface. Solid-surface materials make durable, stylish, and easy care surrounds. Though these smooth acrylic products are sometimes pricey, nothing beats them for ease in cleaning, and the material is resistant to fading and mildew. Even better, solid-surface tub and shower kits can be easy for do-it-yourselfers to install. These kits generally consist of precut panels and curved corner moldings. They are designed to go with standard fixtures, however; nonstandard installations require professional help.

Ceramic, porcelain, or natural stone tile. Thanks to our global marketplace, tile has style and it's played out in countless colors, grains, designs, and materials. Waterproof, durable, and easy to maintain, tile is a logical choice for tub and shower surrounds. One drawback: The grout can mildew, making it difficult to clean. The best advice is to avoid white grout and stick to gray or darker colors. Also, natural stone products will need to be sealed and cleaned properly (each type of stone with its own special requirements), so make certain you select a product you are able and willing to keep up.

Prebonded tile. To make things easier on the home remodeler, some tiles come in prebonded sheets. Small mosaic tiles (which measure about 1 in. sq.) come bonded to sheets of 1-ft. by 1-ft. or 1-ft. by 2-ft. fiber mesh. These sheets go up faster than loose tiles, because you don't have to set each piece individually. Pregrouted sheets of 4-in.-sq. tiles with flexible synthetic grouting are also on the market. Stick the sheets to the substrate surface first; then apply a thin bead of caulk around the edges. As with mosaic-tile sheets, installing pregrouted tile sheets is not as time-consuming as laying loose tile. Even better, the sheets are easier to install.

Fiberglass. Fiberglass is waterproof, durable, and simple to clean. Many companies manufacture three- and five-piece shower/tub surround units in various sizes. Installation isn't difficult if your walls are plumb and have been properly prepared.

Fresh color can make even the simplest shower more inviting. The wide range of tile choices means you can create signature art on your shower walls. These homeowners had fun with aqua mosaic tiles and a fishbowl concept.

custom showers

● ● ● NEW SHOWERS ARE DOING THEIR PART TO TAKE THE BATHROOM FROM UTILITARIAN TO EXPERIENTIAL. Conventional shower stalls and shower-tub combinations are practical and affordable, but if you have the chance to upgrade, you can turn your shower into much more than a place for a quick douse. High-output shower panels, recirculating pumps, custom multihead installations, handheld sprayers, and steam generators are all options for rethinking the basic shower.

Designers have found artful ways to integrate showers into an overall bathroom design. Materials such as glass block, sheets of glass, and half-walls are used to set off a shower space. Sometimes clever tiling—even in the absence of walls—can define an area for showering. Working with tile, stone, or a solid-surface material, such as Corian, a shower installer can create virtually any shape and configuration to fit your available space. Having choices in tile and stone also means you can capture any look you desire, from clean and modern to natural and tactile.

ABOVE A more typical shower and tub duo is designed end to end. Glass shower enclosures are the trend, but require a daily wipe down to look their best.

LEFT There are new shower functions on the market that can turn a daily shower into a lavish experience. These water tiles from Kohler, with their 54-nozzle spray heads, deliver water in a luxurious way. Does it get any better than this? It does if you add a chromatherapy function that adds mood-altering, colored-light sequences.

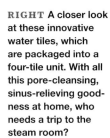

RIGHT A closer look at these innovative water tiles, which are packaged into a four-tile unit. With all this pore-cleansing, sinus-relieving goodness at home, who needs a trip to the steam room?

MULTI-JET SHOWERS

Shower towers and jetted panels can make your shower seem like your own private water park. The current designs actually originated in Europe, where changes in in-wall plumbing are difficult because of the age and type of construction there. Some of the fixtures are attached to the shower wall at only two points and then connected to hot and cold water lines, making them good candidates for a remodel in which an existing shower is still sound.

Panels with a variety of water outlets are available. One model, for example, has seven jets, including adjustable showerheads, four body sprays, and a handheld shower. These models are adjustable to suit your height and desire for placement of body sprays. The intensity of the sprays is also adjustable. Now, water tiles can be inset to simulate rain. Do you prefer your gentle rain at sunset? Then you might want to choose water tiles that offer color options.

All this performance, however, drinks up a lot of water. Federal regulations limit the output of a single showerhead to 2.5 gallons per minute, but because these new showers have multiple heads, the consumption of water can be far greater and may require oversize supply lines (and a dedicated water heater). Many new multi-jet showers have the option of installing a special basin to recycle the water during the massage function.

ABOVE If you're dealing with a small space, go big on creativity. Similar to European-style water closets, this tub and shower combination operates like a single side-by-side unit that is open to other areas of the bath. A series of cedar slats extend the shower floor across the width of the room.

specialty showers—the steam shower

●●● TIME SPENT IN THE STEAM ROOM IS A LUSCIOUS PART OF THE HEALTH CLUB OR SPA EXPERIENCE. Steamy warm air can release muscle tension, loosen joints, and refresh worn-out bodies. Now with manufacturers focused on the home spa market, there is a variety of prefabricated showers that include steam capabilities as well as conversion kits that transform a conventional shower to a vaporous one. Of course, you can also decide to add this amenity along with many others to a newly purchased custom shower.

A steam shower is basically an enclosed area with a vapor-tight door. You can enjoy steam created by your shower, or you can equip your shower with a steam shower generator. The steam generator heats water to a boil and then releases the resulting vapor into your sealed up area. A seat or bench gives you a place to relax while you soak up the steam.

Obviously, water vapor from a steam room can be damaging to your home if not properly controlled, so you must be diligent about waterproofing the shower. Nonporous tile, stone, or synthetics are the best materials to choose. All tiles should be set in a mortar bed, not stuck to water-resistant drywall with adhesive. This should be discussed with your tile contractor.

BELOW Tile is one of the best surfaces for a steam shower. This vintage-look subway tile creates a stylish locker-room look, complete with the steam room feel.

RIGHT Seal up your shower with a vapor-tight enclosure and you've got your own private steam room. This spaciously sized, square shower, with a floor-to-ceiling glass and tile enclosure, gets steamy with every use, unless the door is left ajar.

ABOVE A transom is a handy way to flip back and forth between a steam shower and one that is vented to the ceiling where a fan can pull the steam up and away. The ceramic subway tile and Carrara marble in this shower are completely impervious to water absorption.

LEFT By sealing your shower closed at the bottom and sides, your shower will hold the steam while a vent pulls and releases it up top. Only a small slip of open space remains between the ceiling and the sides of this simple glass shower.

LEFT Find an enclosed nook or niche in your backyard, hook up some water lines, add a faucet, pipes, and an on-demand water heater and you've got the recipe for simple, natural luxury. Here, you can experience this indulgence in a stone and wood setting.

ABOVE On this Iowa farmstead, a tankless water heater means the simple pleasure of an outdoor shower can be enjoyed for multiple seasons. The owners planted roses and other aromatic herbs in and around the shower to enhance the experience.

• specialty showers—the outdoor shower

Fresh air, sunshine, and water make a delightful combination of elements and energy. That's why a day at the pool or the beach is such a lure—and such a rejuvenating whole-body experience. With an outdoor shower, you can bring that invigorating mix to your backyard. An outdoor shower can be as simple as a hose and nozzle tethered to a tree, or an elaborate setup with a curtain and marble floor.

Of course, as those who have tried it know—the only thing better than an outdoor shower is a warm outdoor shower. If you aren't able to hook yours up to your home's hot water system, consider a hot-water-on-demand unit, such as those made by Noritz® or Bosch®. Instead of using energy to keep a big tank of water constantly warm, ultra-hot gas jets heat the water once it's turned on and as it flows through the system. Many homeowners are enjoying the advantages of this type of system inside the house, too, and the energy and money savings it promotes.

TOP A perfect spot to tuck in a romantic outdoor bathroom is just off the master bedroom. Horizontal wood paneling (sealed well with several coats of polyurethane) plus industrial materials cleverly placed as screens, create an open-air retreat.

ABOVE A large, rain-bonnet showerhead increases the diameter of the spray and the enjoyment of this simple outdoor shower setup.

LEFT You don't need anything fanciful to create an alluring outdoor shower. An open-air stall of redwood, with treated decking as the floor, offers an extrasensory shower time.

faucets and fixtures

●●● TUB AND SHOWER FIXTURES HAVE EVOLVED INTO HARD WEARING, high functioning and good looking accents for your bathroom. In fact, there are so many options that shopping angst is a real possibility. One thing to keep in mind is to purchase tub and sink faucets that complement each other in style and match in finish. Polished chrome remains the hands-down favorite because of its bright, clean look and because it keeps its shine without much maintenance. Other options include matte chrome, pewter, and epoxy-coated finishes in a range of colors.

Whether your bathroom project involves a minor spruce-up or a major overhaul, consider the new showerheads on the market. Many save water and money while they give you a more satisfying experience. After all, we've each met up with showers where the water temperature fluctuates erratically, the water flows out in a wimpy drizzle, or the pressure is so hard it feels as if you're in a super-soaker fight. Now is your chance to get your shower flow under control.

Consider faucets and fixtures the jewelry as you outfit your new bathroom. They are important accents to your room's design. With a striking bronze finish, these classic fixtures add visual structure to the bath's neutral palette.

ECO-CONSCIOUS FAUCETS

W ater conservation is sure to be an evergreen topic—one that is already of critical importance in the West and Southwest. Just think about the numbers: According to the U. S. Environmental Protection Agency (EPA), the average American will use 100 gallons of water each day (enough to fill 1,600 drinking glasses); a house full of leaky faucets can drip more than 3,000 gallons of water in a single year; bathroom faucets run an average of two gallons per minute, and to fill a bathtub takes an average of 50 gallons of water or more per soak.

A terrific place to start your water conservation education is by visiting the EPA's WaterSense® program at epa.com/watersense. WaterSense labels products which it has deemed to be water efficient, similar to its sister program for appliance grading, ENERGY STAR®.

Here are a couple of options to decrease your energy and water footprint, while saving money:

Thermostatic valves. These valves add a safety component to your shower, while subtracting energy consumption. Look for systems that lock into a temperature. A high-end example is Kohler's DTV™ series, which uses smart technology to control every aspect of the shower's function, from temperature, to water pressure control, to multiple programmable showerheads (you use them for only the length of time you need).

Beyond low-flow showerheads. Some smaller companies are also creating smart technology that controls warm water usage. Evolve™ showerheads are getting a lot of attention for their patented ShowerStart™ technology, which alerts the user when the shower water is the right temperature by slowing the pressure to a trickle. Pull the handle when you're ready to wash up, and the shower starts up again at the just-right temperature. According to the company, this can save a family up to 2,700 gallons annually.

TOP Shop for new showerheads with low gallons per minute (gpm) flow. In Kohler's new Forte® showerhead, the spray engine delivers a 1.75 gpm, which is estimated to save 30 percent over conventional showerheads.

ABOVE For a shower that's relaxing and gets the job done, include multiple showerheads. Clad in classic subway tile, marble, and glass, this shower's vintage feel is enhanced by simple chrome fixtures. The rain-bonnet showerhead calms, while the handheld shower cleans.

sinks that add splash

●●● A BEAUTIFUL SINK CREATES A DESIGN FOCAL POINT IN YOUR BATHROOM. Fortunately, the sink has gone through the same style upgrade as its big brother, the tub. Vessel sinks, trough sinks, stone sinks, steel sinks—you name the look and the material.

Porcelain (also known as vitreous china) has long been the dominant choice of materials for creating sink bowls. Solid-surface sinks, where the bowl and counter are one piece, have also been widely popular for their endurance and ease of cleanup. But now these favorites are moving over to make room for glass, stainless steel, and solid-surface sinks. Vessel sinks have also popped onto the scene, sitting proudly atop the counter as opposed to their undermounted cousins.

ABOVE Undermount sinks are designed to drop into vanities and to be paired with a water-safe counter-top. This porcelain basin and granite counter is a seamless installation, which makes cleaning easier.

RIGHT Sinks have come out from under the counter to add a happy new design twist in bathroom schemes. In aqua glass reminiscent of sea glass, these dual sinks add a fresh look. Wall-mounted faucets that extend to the center of the basin keep splashing to a minimum.

mount. Narrow your choices by first determining whether you want the classic looks of a pedestal sink or the storage capabilities of a vanity sink. If a vanity type is the winner, then determine whether you want an undermount or vessel sink. Undermount sinks make cleanup easier, with fewer crevices and curves to keep clean. But they cost slightly more, are harder to install, and require a waterproof countertop.

Material. There are more materials to choose from than ever before, each with their own set of qualities (see page 123 for a discussion of materials). New materials are even improving on old favorites. For instance, tempered glass sinks may appear fragile but they can take a beating. In testing, they have proven more impervious to scratching and chipping than porcelain.

Fittings and faucets. Vessel sinks tend to require wall-mounted faucets, which can add cost to your project. Make sure the faucet extends well into the sink and is properly proportioned to the sink's size to avoid splashing outside the sink's perimeter. To make your sink a splash-free zone, the bowl's depth should be at least 12 in. deep.

TOP For Zen-like simplicity, look for natural materials and ultra-spare designs. This wall-mounted system with its stone vessel sink and wooden shelf counter is the ultimate in simple, functional design.

LEFT During the latter part of the Victorian era (the late 1800s), style came to the forefront in apparel and home design. The curvy and stately console sink became all the rage. In this fresh cottage setting, the two-legged beauty is a focal point.

• vanity sinks

Sure they look better than ever, but sinks still have to stand up to hard use. In other words, they have to serve up substance with that style. Sinks dropped into a vanity are popular because they provide storage caches and counter space. What's more, in today's marketplace you can pick from a vast range of cabinet, table, wall-hung, or retrofitted vanities (created from vintage furniture) that will complete your bathroom scheme and your design dreams.

To narrow your focus in this vast array of design opportunity, consider choosing a vanity that visually links to your home's architecture to help maintain design flow throughout your house. You could shop for cabinetry that shares a wood grain with your home's moldings, for instance, or scout out cottage-style or retrofitted cabinets to slip into a bungalow's bath makeover. Now that bathrooms have the opportunity to make a design statement rather than to be simply a home-building afterthought, the choice is certainly yours.

TOP A white-on-white scheme is a clean, classic choice for both new and old home styles. To keep this white room from becoming too sterile, the owners tossed in a vintage stripe, woven area rug. Though a small element, the black frames add yet another visual anchor.

RIGHT When choosing a vanity sink, you have the option of custom designing with separate sink and vanity styles. This clever configuration technically qualifies as a vessel and under-mount sink, thanks to a raised glass counter-top. The glass prevents the deep counter from visually weighing down the modern room.

FAR LEFT In a contemporary Arts and Crafts influenced home, strong, clean lines and warm wood finishes are the order of the day. Stretching from shower to bathtub, this single sink vanity is the room's focal point. It ties this room together and links to the home's strong architecture.

LEFT Richly-appointed vanities make the grade in traditional style houses. In Ithaca, New York, this marble-top vanity and carved sink mimic classic styles, with jewelry drawers on each side. The pewter faucets are a classic and quiet choice.

BELOW In large bathrooms, you can customize a wall of cabinetry to suit your needs. In this St. Louis home, the vanity wall includes a lowered dressing table and a tall cupboard, giving the cabinetry a sophisticated, furniture-like look that is appealing today.

ABOVE Who could deny the clean and sculptural beauty of a pedestal sink? With separate hot- and cold-running faucets and a built-in back-splash, this simple beauty has timeless appeal.

RIGHT When you shop for wall-mounted sinks, you'll find that some have a built-in basin and other styles come in one piece. This marble-topped corner sink with a separate porcelain basin is a petite vintage find. Its short skirt prevents the sink from tak-ing too much visual space like a longer skirt would.

• pedestal, wall-mounted, console, and freestanding sinks

A pedestal or freestanding sink is a one-piece basin and countertop. The classic pedestal design has remained popular through the decades because of its charm and ability to squeeze into the tight spaces of small bathrooms and powder rooms. Pedestal sinks come in modern and classical styles.

Freestanding sinks, with legs and a bit more surface space, are another option with vintage appeal. Small wall-mounted units take up the least space of all, especially if you consider corner units, and slip into the tightest of spots. Like vanities, many pedestal and freestanding sinks are now available in multiple heights, including the standard kitchen counter height of 36 in., making them more comfortable to use.

ABOVE Consoles come in all shapes, materials, and sizes. With its compact size and tapered sides, this slender console lends sculptural beauty to a slender powder room. The freeform glass vessel sink and modern pump-like faucet add to its artfulness.

LEFT Sometimes referred to as "hotel style," chrome, marble, and ceramic tile in rectangular (subway tile), hexagonal, or basket-weave patterns, this clean, classic look hails from the 30s and 40s. A wooden bureau pulled in for storage warms the space with its gleaming grains.

● top-mounted sinks

Above-counter sinks are commonly referred to as "vessel sinks," and hark back to washbasins that rested on bureaus and tables before modern plumbing. Because they are shaped from a large range of materials—from glass to copper to glazed pottery—they carry a wide range of price tags as well. The vessel sink's popularity has taken off like wildfire, mostly because of its dramatic and unique design. When shopping for this type of sink, don't be taken in by good looks alone. Be certain that the material will hold up under daily use and that the basin has enough diameter and depth for comfortable use.

TOP Glass tile, wooden counters, and porcelain basins are popular materials in today's sophisticated bathrooms. With a lever tap that can sit slightly off-center, this shallow but wide basin can sit on a narrow counter for space savings.

RIGHT Vessel, or top-mounted, sinks offer a higher sink height, which many users find more comfortable for washing up. These ample basins with their smooth, square shape bring farmhouse appeal. The gooseneck faucet and double-handled lever faucets are also designed for easy use.

ABOVE Humble materials yield quietly elegant results and create a bathroom as easy on the budget as it is on the eyes. A stainless-steel work table like those found in restaurant supply stores forms a durable counter for a bowl-like porcelain basin. A natural wood divider adds privacy and a place to hang the mirror.

RIGHT Vessel-sink designs are artful additions to bathroom design. Make a modern statement with stainless basins and pump-like, single-handled faucets that are reminiscent of old farm pump levers.

MATERIAL MATTERS: WHAT TYPE OF BASIN TO BUY

Vitreous china. Pedestal, wall-mounted, and undermount sinks made from vitreous china, or porcelain, are virtually impervious to any type of cleanser that you can douse them with. They can, however, chip and crack over time.

Ceramic. Artful sinks spun out of glazed ceramic can be fairly durable, but they are more prone to chipping and cracking than vitreous china.

Enameled cast iron. Like their cousin the clawfoot, cast iron sinks are tough and resistant to cracking. The finish can chip if hit hard enough.

Stainless steel. Now more common in residential bathrooms, these durable sinks are easy to clean and hide dirt well. Other metals such as pewter, nickel, or silver plate are softer and require careful upkeep.

Cultured stone. Created by mixing crushed stone (often marble or granite) with polyester resins, some inexpensive cultured-stone sinks have a gel-coated finish to give the sink its color and texture. This gel-coat will fade, crack and blister over time. More expensive cast-polymer sinks with a higher stone or mineral content are more durable.

Solid-surface materials. Sinks made from Corian, Silestone® or other similar solid-surface materials are durable. Scratches and stains can be buffed out of these synthetics, and they can be fabricated into one-piece countertops that are seamless and easily maintained.

faucets

●●● THERE IS AMAZING DIVERSITY IN BATHROOM FAUCETS. You will pick from a variety of finishes, styles, and types. Two common pieces of advice to heed as you shop through the maze of options: Don't be taken in by an inexpensive faucet, and do buy your fixture and faucet at the same time so you can determine their compatibility. Though a well-designed faucet can be inexpensive, quality parts mean the difference between a long-lasting, non-dripping purchase and one that wears out quickly.

In fact, a well-made faucet should come with a lifetime warranty against drips and parts failure. You can also expect a quality valve within that faucet which holds your water temperature steady and prevents it from changing abruptly as you shift water temperature. One quick test for solid construction as you're shopping the aisles of your local retailer: Faucets of lesser integrity tend to be lightweight because they contain more plastic.

ABOVE Single-lever faucets are a popular choice for vessel sinks because of their simplicity, and because both sink and faucet are clever updates of early versions of home water carriers: the pump and basin. In polished chrome and extending from a polished chrome console, this combination is compact and visually seamless.

more about...
FAUCET CATEGORIES

Centerset. This category includes one- or two-handled styles where the valves and spout are located on a shared base. One-handled models are easiest to adjust because a single lever controls both the hot and cold water lines; two-handled models have to be adjusted for temperature separately.

Widespread. These faucets have a spout and two valves, all mounted separately, from 8-in. to 20-in. apart. They're typically an upgrade from centerset models.

Mini-widespread. These faucets are similar to widespread, but they aren't so wide—the center-to-center distance between valves is 4 in.

Wall-mounts. This type of faucet is mounted directly to the wall, making them a great companion for above-counter sinks. They have long spouts to reach to the center of the basin. Though these faucets offer a sleek, built-in look, be prepared for the extra cost in installation. Plus, your plumber will need to know the exact height and location of the faucet early in the building or remodeling process.

TOP One key to a well-fitted faucet is whether or not water is delivered close to the center of your sink. This is especially true for shallow vessel or trough sinks, such as this one-piece stone, trough basin.

LEFT A waterfall faucet creates a flow that is as special as the design of this modern beauty. Wall-mounted faucets such as this keep plumbing out of sight behind closed walls.

FACING PAGE Mini-widespread faucets are well sized for small sinks. With its exposed chrome legs and plumbing, this dual vanity has faucets that extend the industrial-chic look—the classic cross handles.

toilets and bidets

● ● ● THOUGH AT FIRST GLANCE IT MAY NOT SEEM THAT YOU HAVE MUCH CHOICE when it comes to selecting a toilet, your decision will be affected by the look, the maintenance, and the function of today's commodes.

There are one- and two-piece toilets. The two-piece toilets, with a separate bowl and tank, pose more tough-to-clean seams. One-piece units come out of the box in a single unit, and often sport a lower, more modern profile. Though there is not much, if any, difference in the function of these two types, the one-piece units are generally higher in price.

Other new designs feature behind-the-wall tanks for sleek, bowl-only looks. You might also want a toilet that doubles as a bidet with its built-in washing systems. Or, consider a tank and seat with self-cleaning ions.

Fanciful detailing allows some toilets to feel at home in bathrooms appointed with ornate, traditional styling. The designer of this St. Louis bath took advantage of porcelain's paintable surface and had the wallpaper's motif custom painted on the tank.

This modern guest bathroom features a divider wall that adds design interest and a sense of privacy. The newly created "stall" features a hardworking Toto® toilet, known for its quiet, high performance flush.

There are many new bells and whistles sounding on the latest toilets, including self-closing lids and surfaces that repel bacteria. This Compact Cadet® model from American Standard offers those options, plus water savings and flushing power.

Low-profile toilets take up less visual space in a bath, but offer the same seat comforts. The San Raphael by Kohler sits just over 21 in. tall for easier lift off. And it has serious flushing power with water savings.

WHICH FLUSHING SYSTEM IS RIGHT FOR YOU?

pressure. This type is considered the best for families where the toilet sees some hard wear. Pressure-assisted toilets create the most flushing power, but for that you will likely have to put up with a louder flush. Some manufacturers are offering quieter, more water-efficient power systems, such as with Kohler's Power Lite™ system. You also need to make certain that your home has the water pressure it takes to adequately supply this system. Call your water supplier or check your pressure with an inexpensive gauge that connects to an outdoor spigot. You'll need at least 25 pounds per square inch (psi) for the toilet. If you need to adjust your water pressure, don't go above 80 psi, which can harm toilets and other fixtures.

Vacuum. This flushing system is best for bathrooms where silence is a virtue. Look for the models that promote the most flushing power, however, as performance of these models is weaker than pressure–assisted flushing, even though they come with a similar price tag.

Gravity. Shop for this system if you want a quiet flush or have low water pressure. Beware of low-price models that may not be up to the job your family requires. Gravity toilets rely on a flush valve to discharge water from the tank and into the bowl. Valves 3 in. to 3½ in. wide help deliver more oomph than gravity models with 2-in. valves. Ask to see the manufacturer's specifications for the flush valve when considering a gravity toilet.

• the latest toilet trends

Commodes today are more thoughtful in many ways. Consider the comfort height models, which raise the rim of the toilet from the usual 14-in. height to as much as 17 in. off the ground (which is close to the standard seat height of 18 in.). The added height makes it easier to get on and off.

More and more efficiency is a happy result (from the consumer's point of view) of the competition between manufacturers who are attempting to beat the 1.6 gallons per flush legal standard set by the government in 1994. Some models with dual-flush technologies use a mere 0.8 gallons to clear liquid waste.

BELOW Looking for a bit more luxury in a loo? This sleek Fountainhead™ design from Kohler offers integrated bowl lighting and a heated ring. With just a slight touch the lid opens and shuts.

ABOVE Slender, elongated models slide discreetly into narrow spaces. This low-rise, one-piece model has a rounded tank design for a soft look.

Toilets seem to be the next bathroom fixture in line to undergo big changes in both design and function. The Purist® Hatbox® design from Kohler equalizes form and function in one streamlined design. This high-end unit has a taller seat for comfort and a self-closing lid.

129

• special spots—bidets and urinals

Bidets have long been a commonplace feature in European bathrooms, where daily bathing is not the norm. In the United States, where we love our tub and shower experiences on a daily basis, this fixture has been slow to catch on. Even so, they offer obvious hygienic benefits, as well as obvious water savings. Today, bidets come as freestanding units that typically sit next to the toilet, or more and more commonly, as an added feature on today's new toilets and toilet seats.

URINALS IN THE HOME

a urinal in a private home is seen even less often than the bidet in U.S. households. Still, you may want to check out the models available today, many of which feature waterless waste clearing. According to manufacturers, the waterless urinal can save thousands of gallons of water per fixture each year. More compact and sculptural in design, today's advanced urinals offer a splash-free surface, along with odorless, easy-to-maintain performance.

LEFT Does your nose wrinkle up just a little bit when you hear the word "urinal"? Half the population may not be too familiar with the fixture, but manufacturers are making it easier to get used to with virtually maintenance-free, waterless urinals. This high-end version from Kohler is estimated to save 40,000 gallons of water a year.

LEFT Europeans have long been able to work a bidet into even a small bath space. Here, a modern bidet and toilet sit side by side without spoiling the simple serenity in this London bathroom.

FACING PAGE Bidets are available for the U.S. market in many styles, from freestanding to bidet-toilet combinations. This angular, traditional design is as stately as a sculpture, so it fits right into a sophisticated scheme.

cabinetry and other storage

• • • •

NO MATTER WHAT SIZE SPACE YOU'RE DEALING WITH, PLANNING FOR the right storage elements will mean an easier morning routine. Even more, the surfaces you select for your cabinetry and countertops can impact the final look in a big way—just think of the amount of visual space these elements take up. Careful consideration here will help you create a bathroom that serves you and your family well, and does it with style.

When planning your bathroom's storage components, give some thought to all the items that will be called into service by the members of the household who will be using the room. From hair care products to medicines, from dental products to cosmetics, and from shaving to showering—this may be the most product-laden room in the house, even if it is also the smallest one.

In this chapter, we'll explore the many options for your room's cabinets, countertops, and storage, including some creative and stylish stow-away options that will work for you regardless of whether your room is grandly proportioned or pint-sized in scale.

Well-placed storage is a virtue in any size bath. Wall units and under-counter drawers can make life much easier. Though these homeowners wanted the lighter look of under-counter space, they tucked in a handy built-in drawer unit.

cabinetry

••• CABINETRY PLAYS TWO BIG ROLES IN YOUR BATHROOM'S DESIGN. The most obvious one is its function as a stash-all. But also consider how the style and finish of the cabinetry you select can take your room in many directions. You can select polished traditional woods, sleek modern laminates, or the age-worn finishes of rustic country. Though you don't want to make a radical departure from other finishes in your home, the look of your cabinetry is an opportunity to shift into a fresher style, or one that more closely appeals to your aesthetic.

Working enough storage into your limited space can seem like a puzzle—and it is! But again, you have many options from which to choose, from under-sink cabinets to ceiling-high linen closets. You also have choices in fittings. Vanities and linen closets can be fitted with drawers, shelves, tilt-out hampers, wire baskets, and other specialized accessories (How about a warming drawer for your towels?) to maximize their capacity and function. Don't forget about the classic medicine cabinet, whose shallow shelves are perfect for keeping medicines, creams, dental supplies, and other small items in view and within reach.

ABOVE When built-in storage isn't practical, choose from the wide selection of ready-made storage units. This slender cabinet, in a light walnut finish, fits between the sink and shower—a very handy spot.

RIGHT With the right planning, storage units can do even more than provide storage. These custom-made units also act as room dividers between bedroom and bath. And the furniture-style built-in gets a lighter look simply by being a foot shorter than the ceiling.

TOP Use awkward niches to your advantage. With its pull-out hamper and towel-size shelves, this slender built-in unit fits into a space left by a retrofitted vent in an older home.

ABOVE Plan for storage early in the building or remodeling process and you can tuck it into walls, giving you more space. Against a mosaic of watery blue glass tile, this small cupboard fits between the studs and adds storage and contrast to the space.

LEFT Another option is to build out from the wall for the storage you need. Borrowing an idea from the pantry, a door fitted with small shelves swings out for easy searching in this attic space.

•the elements of a quality cabinet

As with many other purchases you'll be making for your bathroom project, when you look for stock cabinetry it's advisable to select the best your budget will allow. You'll bump into many terms that give you clues to the quality and durability of cabinets as you begin shopping. Once you arrive at the right style, check the manufacturer's specification sheet against the list on the facing page.

Quality cabinetry can become the focal point of your bathroom design. With drawers that bow slightly and artful hardware and moldings, this vanity has all the appeal of a fine furniture piece.

TOP When shopping for vanities or cupboards with drawers, one of the best quality checks is in the drawers. Look for dovetail construction, as revealed by this Shaker-style vanity.

ABOVE The right cabinetry can enhance your bathroom's style. In this seashell-studded bathroom, creamy-colored cabinetry plays a quiet, supporting role to let natural accents be the stars.

CABINET QUALITY CHECKLIST

t o be assured you're buying cabinetry that will endure the daily bathroom routine, check all cabinetry against this quality checklist:

1. Look for manufactured cabinets with a baked-on finish called catalyzed conversion varnish, which bath designers feel is the main reason to buy manufactured cabinetry.

2. Two door cabinets up to 36 in. wide can use butt doors. Budget cabinets use a center divider (called a center stile) to give them more structure. A center stile can also be used to hold up skimpy shelves.

3. Shop for 3/4-in. shelves in 65-lb. particleboard for the most stability. Plywood with a vinyl or wood veneer is also acceptable. Particleboard is the most popular choice these days because it is less likely to warp than plywood.

4. When buying a frameless cabinet, look for a minimum of 3/4-in. melamine-covered 65-lb. particleboard construction for all horizontal pieces, such as shelves. For vertical sections, make sure the product's specs reads 5/8 in. or better.

5. Drawers should also be a minimum of 5/8-in.-thick solid wood, preferably with dovetail construction. Be sure to measure the interior size on drawers. Substandard drawers will be shorter in depth than full depth of the unit and is a sign you will want to move on.

6. Look for specs that indicate a minimum of 100 lb. capacity, or check for bottom-mounted drawer slides. If there are two for each drawer, the drawer will operate more smoothly.

7. Solid hardwood or semi-hardwood drawers should be finished with a catalyzed conversion varnish. Check to be sure this varnish covers all surfaces. Cabinet interiors also should be finished with this baked-on sealer, or be covered with heavy vinyl wrap or a wipe-clean interior.

8. Quality cabinetmakers will offer a lifetime warranty on all hinges, door slides, and other mechanical or moving parts.

A walk-in closet or dressing area placed
adjacent to a master bath not only makes
getting out the door easier, it offers a nearby
stash for bathroom extras. In this modern
space, a frosted-glass vanity links to the
glass-block shower wall and offers even
more storage wells.

LEFT Use partial walls to your advantage to create niches for shelves or closed storage. Check out the storage opportunities provided by this small wall, designed to be flush with the vanity. It holds towels and provides a place to plug in a small flat-screen television.

BELOW Built-in storage cabinets not only enhance the stowing power in a bathroom, but reinforce the design as well. The built-in storage in this farmhouse bath recalls the simple beauty of the Shaker style, and creates an extra closet full of space in an awkward niche.

BOTTOM Borrowing the clean, commercial look found in medical cabinets, cabinet manufacturers have created industrial-chic style. This slender glass and metal cabinet holds cosmetics and corrals favorite collectibles for a personal touch.

BELOW
When refinishing spaces, new walls can create excellent opportunities for building shelves in handy places. This clever homeowner makes even more use of potential dead air space to install a space-saving, three-sided vanity.

RIGHT Painted a creamy-white and freshened with shiny stainless pulls and knobs, this bathroom built-in, once a hallway linen closet, is now more conveniently located to serve the bath.

FAR RIGHT Great storage is sometimes less about the amount you have than where you put it. A simple and sleek cube of storage keeps vivid towels close at hand and provides a perch for happy mementoes.

BELOW Borrow a page from kitchen storage and take your cabinetry from floor to ceiling. This hardworking wall of bathroom storage doesn't feel imposing because the cabinetry is clean and white—with some open storage.

LEFT Get personal with your storage caches. Mellow silver of all types and shapes makes an intriguing collection, and the vessels create brilliant hideaways for sundries and cosmetics.

BOTTOM LEFT When using open storage, the trick to keeping things neat and tidy is to fill the space with one type of item. Here, towels add another black accent and draw the eye to the dark line that rims the upper shower.

BELOW Built-ins today are designed like beautifully-detailed pieces of furniture. With its crown molding, decorative glass, and raised-panel detailing, this slender linen closet is pretty enough for the dining room.

• achieving the look

You have basically three choices when it comes to fitting your bathroom with cabinetry: custom, semicustom, and stock.

Custom cabinets are made to order, which means you can get them in any size and shape. Local cabinet shops are often the source for custom cabinets as well as the many talented carpenters who work on their own. If you have contracted for custom cabinetry, be sure to check the portfolio and references of your cabinetmaker, as this is an area that requires skill. Though you'll get the exact cabinetry you want, you will also pay a premium price.

Stock cabinetry, like the many other elements you will choose for your project, now comes in many styles and finishes. You can find cabinetry on hand at home improvement centers and kitchen and bath showrooms, and often they can be picked up and installed in the same day. These cabinets are mass-produced using standard sizes and shapes.

You might also aim somewhere in the middle of custom and stock cabinetry by considering semicustom cabinets. In the most basic terms, semicustom signifies stock cabinetry that can be altered in some way to better meet the consumer's needs. For instance, there are more choices in finishes, design, and component options. You can also find built-to-order or semicustom cabinetry made by large manufacturers at specialty bath shops or home improvement stores, but delivery may take longer than for stock.

When you are in the market for stock or semicustom cabinetry, your best bet is to measure carefully the area you've reserved for storage and shop with your measurements close at hand.

Custom cabinetry can be designed to fit your storage and style needs. A mirrored medicine chest that stretches across the length of the vanity adds sparkle to the fresh cottage style.

gallery

cabinet door styles

Cabinets get their looks mainly from the style of their door and drawer fronts. These basic door styles—and the numerous variations on them shown in the photographs throughout this book—are available for both frameless and face-frame cabinetry.

Frame and Flat Panel (or Mirror)

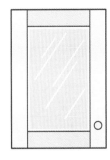 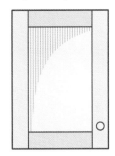

Raised Panel

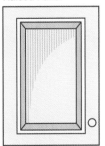

Curved Raised Panel

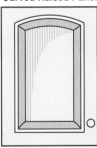

Mullion Glass Panel (or Mirror)

Cathedral Raised Panel

Beadboard

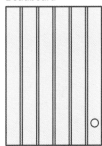

The variety of styles available in readymade cabinetry seemingly has no end. This modern, long, and low credenza-style vanity brings living room style to the bathroom.

Multiple wood finishes can give new custom cabinetry a look that appears to have evolved over time. Featuring wood and painted finishes, this dressing area sports a vintage look complete with romantic sconces and a vessel sink shaped like a vintage bowl.

creative
storage

●●● SOME HOMEOWNERS AND DESIGNERS REPURPOSE VINTAGE OR UNFINISHED FURNITURE PIECES AS HIGHLY PERSONALIZED VANITIES.

Often, converting a bureau or chest to a working vanity is a relatively easy project, especially if you choose a free-standing vessel sink and wall-mounted plumbing fixtures. Or, for a readymade one similar to this do-it-yourself project, look for the many manufactured choices in freestanding vanities. There are vanities that at first glance look like traditional, bow-front dressers, for instance, or Shaker-style tables.

Clever open storage is an option if you aren't shy about parading some of your bath supplies. Open shelving, baskets hung on walls or tucked under counters, and even a line of Shaker pegs or hooks can hold what you need, especially if floor space is limited.

ABOVE Think of storage as an opportunity to get creative. A salvaged cupboard door and a bright yellow bin add personality and cut clutter.

RIGHT A vanity fashioned to look like a bureau can be a design bridge from bathroom to bedroom. This hefty vanity has cottage-y appeal, and creates a nook for a window seat/storage well.

gallery

cabinet hardware

Pulls and knobs act like the jewelry of your newly-outfitted bathroom. Now cast in many colorful, sleek, and fun shapes, cabinet hardware is a final touch that reinforces your style statement and adds some personal sparkle to the room. Metal pulls come in a variety of shapes from the new organic styles to traditional cup handles. These days, glass pulls are widely available in new colors, shapes, and material combinations, fueling resurgence in their popularity. Novelty knobs and pulls with a spark of fun abound in the marketplace as well.

Metal and Wood Pulls

Glass Knobs

Organic and Traditional Cup Handles

Novelty Pulls

BELOW Design your storage right where you need it for a smoother morning routine. Within arm's reach of the tub, this slender, tiled recess keeps bath supplies organized and ready.

LEFT When in the planning stages, work as much storage into your floor plan as space allows. The shirred fabric door panels on this bump-out linen closet stash towels and sundries while adding softness to the otherwise slick surroundings.

countertops

● ● ● WHETHER THEY STRETCH ACROSS A WIDE EXPANSE OR COVER ONLY A LIMITED SPACE ON

a small vanity, countertops play an important supporting role in your room's design. Like the cabinetry you choose, countertops make a visual statement in your room and are an element that you're not likely to change very often. Consider materials that are neutral in color and limited in pattern so you won't tire of the decor as quickly.

The counter's main role is to be sturdy enough to stand up to moisture. When choosing the look you like, consider the surface's resistance to stains, water, and wear. There is a wealth of materials that meet the bathroom's requirements, including new and creative options. You'll choose from a variety of synthetic and natural surfacing products available in a range of colors, textures, and price ranges.

Loved for its subtle graining and high-class reputation, marble remains a favorite choice for bathroom counters. When properly sealed, as in this shades-of-gray bathroom, marble is also a durable choice. You can rest assured this material won't go out of style.

By selecting a one-piece, seamless counter surface and an undermount sink, you will make cleaning time easier on yourself. This solid-surface counter has few seams and crevices for grime to hide in.

Designing with a mix of surface materials can bring textural interest to an otherwise spare bathroom. The twilight blue laminate cupboards create a strong focal point when contrasted against a marble counter.

Natural stone counters will have more seams and texture than the solid-surface synthetic look-alikes. The owner of this limestone-clad bathroom prefers the imperfections and patina of the real thing.

• natural materials

Your natural choices include stone, tile, wood, and concrete. Of the earth-based options, stone will be the most expensive choice. On the plus side, stone can be cut into one of the most durable and beautiful surfaces which, if sealed properly, will last a lifetime with little maintenance. Many homeowners choose stone for smaller applications where beauty is a virtue, such as in a powder room or small guest bath.

Tile is a category that has exploded over the last few years. There are artful and handcrafted tiles on the market that are carved out of ceramic, glass, and porcelain. The flexibility in design and use of this classic counter and backsplash covering means you can make a personal and colorful statement. Plus, designer and handmade tiles are more available than ever at home improvement stores and bath shops.

Wood, when properly sealed against moisture, becomes a warm and workable surface. Think about the type of woods used on a boat for the best bets. These include marine grade plywood, cedar, and teak.

The popularity of concrete as a counter material waxes and wanes, but because it can be cast into any shape and dyed any color, it remains an intriguing choice. Whether or not you choose it may depend on whether there is a good fabricator in your community.

FACING PAGE
Thanks to the interest in all things natural, wood has become a popular choice for counters. Like other porous materials such as marble, slate, and terra-cotta, wood needs to be properly sealed against moisture. This space-saving corner sink has a protective coat of polyurethane.

LEFT Travertine has appeal due to its subtle graining and lightly-textured surface. This one-piece travertine, trough-style sink and counter is not only as hard as a rock, it has the advantage of easy cleaning because there are no seams and edges to trap grime.

ABOVE Polished to a smooth, high sheen, this black marble slab with its hand-chiseled edge looks as if it came straight from the quarry, but is actually sealed and ready to last a lifetime with minimal care.

LEFT Tile is the classic choice for bathroom surfaces, where they are applied on walls, floors, and countertops. In this bath, tile covers all three, plus a custom vanity. Keep in mind that sealed grout looks great longer because it's moisture-free.

countertop choices

Countertop choices come in a range of materials, colors, styles, and prices. At the most affordable end of the spectrum, laminates offer bright colors and can mimic a variety of more expensive materials. Glass tile, often sold as 1-in. by 1-in. mosaics, are now available in larger tiles, more suitable for counter use because there is less grout to clean. Stone tile comes polished or unpolished. Generally, polished stone is easier to maintain while unpolished offers a softer, more natural look. Solid-surface and stone composite countertops like Corian and Silestone are not inexpensive, but offer durability and are also easy to care for.

1 Bright laminates

2 Wood-look laminates

3 Glass mosaic tile

4 Polished stone

5 Unpolished stone

6 Richly-colored solid-surface

7 Spa-colored solid-surface

8 Stone composites

1

2

5

6

3

4

7

8

•synthetic countertop options

Synthetic surfaces include laminates, solid surface materials (DuPont™ Corian is the best-known name, but there are others), and stone composites (such as Silestone).

The least expensive countertop material, by a considerable margin, is still plastic laminate, but that won't make your selection process any less complex. The four major manufacturers—Wilsonart®, Formica®, Nevamar®, and Pionite®—each offer hundreds of colors and patterns.

Solid-surface materials are generally cast of pure acrylic or an acrylic-polyester mix, and can be fabricated on site almost like wood. These products are also available in a wide range of colors, many with added flecks for a bit of texture. Other types approach the look of real stone. The great advantage of solid surfacing is the seamlessness (you can even get a one-piece counter with a basin molded in) for easy cleaning. This material is virtually stain resistant, but it will scratch and burn (don't forget to turn off that straightening iron!).

Stone composites have gained popularity because they have the look of stone at less cost. The composites include a small amount of acrylic resin or epoxy for increased scratch and stain resistance. Many of the new composites enjoy the eco-friendly advantage of having recycled glass worked into the mix.

Solid-surface counters are available with sinks molded into the single piece. Without seams, the look is clean and simple and stays that way with little attention.

Composite counters, also called engineered counters, are made by blending acrylic polymers (basically plastic) with stone or glass materials. The liquid mix is then poured into a mold. The result in this bathroom is a stone counter look, but in a watery blue shade that will be easier to clean.

Laminate counters have multiple benefits, including their low-cost, easy installation, and breadth of design choices. This modern bath takes advantage of laminate's vast color and fabrication options in this custom red vanity and storage wall.

• keeping your counter clean

With so many counter options and so little time, perhaps your best choice should be based on how easy the counter is to maintain once it's installed in your bathroom. If your bathroom sees heavy daily use and your housekeeping time is limited, steer yourself toward hard-wearing and durable engineered products. If you want a high-style look in a limited space, stone might be your first choice. No matter your preferences, you'll find that each surface has advantages and disadvantages.

Polished granite is one tough counter material. The most important question to ask before you purchase is whether or not the stone surface is sealed against damage. This black granite hides stains, but not dust.

Concrete can mimic the look of stone at less cost, but needs to be sealed and resealed often, as do other porous stone materials. These types of counters are best used in low-traffic bathrooms where looks count.

Limestone has a soft, natural look that is gaining in popularity. In this bathroom, it is paired with stainless steel for a modern design statement.

Though a beveled edge will add dollars to a stone counter, the finished look may well be worth it to you. Also, these edges are safer for families with young children.

Solid-surface materials are a good choice for a family bath that sees some hard knocks. The material is impervious to stains and can be given a good scrubbing—so why not go for white?

CARE AND UPKEEP OF COUNTERTOPS

how a particular material looks is obviously a big factor in deciding on a countertop material. But also consider how it lives. Here are some characteristics to think about:

GRANITE AND MARBLE

Granite is the toughest of your natural stone options, but it must be sealed properly and cleaned daily with mild soap and water, and weekly with a granite cleaner. Sealed marble resists most stains, but the stone can be more easily chipped and scratched. Granite and marble should be resealed annually, depending on the finish.

LIMESTONE AND SOAPSTONE

These porous stones are prone to stains (even when sealed) and scratching. They are recommended for bathrooms that don't see hard wear. Like granite and marble, they can be wiped with mild soap and water, but clean weekly with a recommended stone cleaner.

CERAMIC TILE

This is a great choice for those who want a variety of design options at a relatively low cost. But some tile can stain if not sealed properly at the manufacturing stage. For easier upkeep, make sure your grout is sealed or it will absorb stains and discolor.

LAMINATE

This material can now mimic high-end looks at an affordable price. Laminates are easily scratched and not repairable when scratches happen. Most products show seams, which means soil can get trapped in them. Cleaning is best done with soap and water; stains can be lifted with a slightly abrasive cleaner (containing no bleach). The biggest danger is that flooding the counter with water could allow moisture into the seams, which will eventually cause the material to lift and separate.

ENGINEERED STONE

These products are recommended for high-traffic baths because they needn't be sealed and can be cleaned with a heavier hand. Though the color options are greater, the uniform look of the grain isn't for those who like it natural.

SOLID-SURFACE MATERIALS

Though the cost of this material can rival granite, it offers design versatility and no seams. This product must be installed by a licensed fabricator. Cleanup is worry-free—you can even use abrasive cleaners. Stains, nicks, and scratches can be repaired.

**One final note: Hair products, perfumes, colognes, nail products, creams, lotions, and potions can stain or damage the surface or etch a polished surface. The best way to protect your countertops against these harsh products is to place them on a decorative tray to avoid rings, and use them with care around easy-stain materials.

pairing faucets with counters

●●● WHEN YOU START SHOPPING FOR COUNTERTOPS AND VANITIES, you will need to keep in mind the style of sink and faucets you prefer. These elements should be considered as a unit.

Faucets take on a host of sculptural, sleek, and classic forms, but the excellence of the inner workings and outer finish are the real determinants of its quality. In faucets, the value comes in buying one that performs for a long time without leaking—and that requires durable parts.

TOP Many sinks come with holes drilled for a certain kind of faucet. In this case, a mini-spread faucet fills the bill. Its matte chrome finish is easy to clean.

ABOVE A basin with no holes for faucet fittings allows you to select counter- or wall-mounted faucets. This widespread faucet has a vintage flavor with its traditional cross handles.

RIGHT Vessel sinks can also be partially inset, a modern hybrid of the drop-in sink and top-mounted vessel sink. This homeowner picked an equally clean and modern goose-neck faucet with lever handles.

Vessel sinks have exploded into the marketplace because of the unique designs they offer. They work well with both wall-mounted and counter-mounted faucets. The trick is to make sure there is adequate clearance between the rim of the basin and the faucet, as seen in this wall-mounted application.

faucet styles

It's important to look for a faucet that gets along well with your sink choice, both in function and fashion. Look at these sink-faucet pairings to get an idea of the many hook-ups you can create. Remember that wall-mounted faucets are more difficult to install unless you're starting from scratch or have existing wall-mounted plumbing lines. The best advice when selecting a faucet is to purchase it from the same source as your sink.

1 This widespread faucet is the most common type available today. Note how it is mounted directly to the counter which is predrilled for this standard faucet.

2 Years ago, wall-mounted faucets were the norm. Now, they are making a comeback because they look great with the newly popular vessel sinks. If you are installing wall-mounted faucets, you must first talk with a contractor to determine if you have space in your walls to hold the plumbing lines.

3 A single-lever faucet is the easiest to use and often requires a single hole for installation. This compact fixture features a lever made of elegant horn.

4 Who says the faucet has to be confined to the center of the sink? This partially recessed trough-style sink has one of these centerset faucets at either corner. This type of faucet requires drilling only one hole for easier installation in custom sinks, such as this stone sink.

5 When selecting a faucet to work with a vessel sink, choose one that will point the water flow to the center of the drain. This wall-mounted model has a long faucet that clears the rim and directs the water flow to the drain.

6 The right faucet can be an artful accent on its own. This gracefully arcing faucet and classic cross handles keep the look classic and elegant.

1

2

3

4

5

6

choosing the right finishes

6

● ● ●

A ROOM'S WALLS, FLOOR, AND WINDOW COVERINGS ARE LIKE THE wrapping on a gift. These finishing touches surround the area and virtually seal the look of the space. Just like other design decisions you've made for this hardworking room, your surface materials have to look good while they work to ward off moisture damage, prevent slipping, and keep the space sanitary.

Thankfully, manufacturers of laminates, tile, stone, paint, fabric, and wall coverings have made great strides recently in both the design and engineering of their products. Start shopping for tile and you may feel like a kid in a candy store given all the yummy colors and patterns in ceramic, porcelain, stone, and glass. Moisture-resistant fabrics and wall coverings offer a soft and colorful counterpoint to what can be a room of cold and sterile fixtures, just as stone, ceramic, and even wood materials serve up texture and richness to counter the slick side of the space.

When deciding on your bathroom's windows, walls, floors, and doors, consider budget, style, and durability. In this master bathroom, softly-colored paint finishes and limited fabrics are easy on the budget and the eyes, while the marble floor represents a nicely durable splurge.

LEFT Though wallpaper seems a risky choice in bathrooms, it's actually a fine one if the paper is installed correctly and the room is well-vented. In this charming California bath, an open window is the best defense against moisture problems.

ABOVE Staying in neutral territory is a good idea for those who like a calming atmosphere. In subtle shades of gray and white, this space can easily be reaccessorized for more color and is less likely to go out of style.

• budgeting for backgrounds

There is a wide range of price tags for background materials, so you may want to keep your budget firmly in mind as you search for the right elements. Plastic, laminate, and vinyl come in a variety of price ranges, depending on the quality of the materials, but are generally the most cost-efficient of the background products. Ceramic tile, solid-surface synthetics, and concrete fall into the middle to high-end range of the price spectrum. Natural stone such as granite, marble, or slate, are the costliest. When making your decision, speak with your contractor about the costs of installing each material and factor that into your project's

budget. Fabrics and wallpapers run the gamut in style and price, and many are made especially for moisture-prone areas.

An important point to keep in mind as you shop is the length of time you'll be living with these backgrounds. While paint is a fickle decorator's best friend, it's not so practical to rip up and start again on tile or other background materials. That doesn't mean you need to go for the blandest options. But you should consider how adaptable your choices are to style shifts. If you are remodeling to erase outdated features, such as lilac tile or heavily-patterned wallpaper, it's likely no one needs to remind you of that.

ABOVE Two of the most water-resistant materials around are glass and porcelain. The fluted glass doors on this cupboard will last as long as the variety of tile that mixes and matches in this vintage bath.

LEFT Honed travertine tile is a popular choice in floor and wall tile for bathrooms because it is less slippery, wears like iron, and offers a light, sueded look that goes with anything. The blue green granite and clear blue walls add a fresh contrast to the beige.

tile

●●● CERAMIC TILE HAS LONG WON THE POPULARITY CONTEST when it comes to picking materials for bathroom walls and floors. First and foremost, it's impervious to water. And it's manufactured in a variety of colors, textures, sizes, and patterns, giving you the opportunity to stamp the space with your own personal mark. As a style bonus, tile lends the room an upscale look, but is available at a variety of price levels. Even better, tile can be installed over some existing floors and in older homes where surfaces aren't exactly square or level. (However, tile is an inflexible material that will crack if installed on a floor with even a small amount of wobble.) Once the tile is in place, it requires very little maintenance, especially if your grout is sealed properly against staining.

But of all tile's many attributes, its style versatility makes it a winning choice. Tile can be affixed to floors, walls, countertops, and showers. You can find tile in sizes from 1-in. mosaics to 24-in. square, rectangular, octagonal, or hexagonal shapes. You can use large, light-colored tile to make a small space appear larger, or set tile on a diagonal to make a boxy space more dynamic. High-priced hand painted or engraved tile can be used as artful accents in a field of standard tile. When it comes to tile, you are limited less by your budget than by your imagination.

Ceramic tile comes in all shapes, colors, and types. Getting the grout right is an important factor in how well your tile holds up. In this bath, which features mosaic, hexagonal, and square tile, the grout started out gray and has been sealed—two smart choices.

TOP One of the many beautiful things about mosaic tile is that it comes in prebonded sections, which makes installation easier. Use pattern in small areas that don't see heavy use so you won't tire of it.

ABOVE For many, unglazed porcelain or ceramic tile has a natural, hand-honed quality that is hard to beat. If you envision going beyond a simple pattern, it's best to work with a designer, as the homeowner did in this Boston-area bathroom.

LEFT A glossy finish indicates glazed tile, which has been given an extra layer of liquid glass. Though the choice between matte and glossy is a personal one, consider that glazed tile is more slippery and the finish can scratch. But you can't beat its beauty and water resistance, as the owners of this vintage bath can attest.

•tile finishes

When shopping for tile—especially those intended for the floor, tub surround, or shower—give careful consideration to the finish. Glazed tile—in matte and polished finishes—is the right choice for your bath because it is coated with a sealant that is impervious to water. An unglazed variety will need a sealant to keep it looking good over time. Highly-polished tile is durable, but slippery when wet, and a poor choice for flooring, unless you want to cover it with an area rug. Instead, make sure the product you choose is a soft-glazed type intended for a floor application. If you're not sure, ask to see the manufacturer's slip resistance rating. Tile intended for countertops should be glazed against stains.

One more piece of advice when it comes to selecting tile: Buy extra in case you need to replace individual pieces down the road when the color and pattern may no longer be available.

ABOVE Pebble tiles are easier to install than they look. Many natural pebble tiles come on mesh squares that interlock for easy installation. This accent strip of natural river rock adds texture and a harmonious hint of nature.

ABOVE Glazed ceramic tile offers more color options than unglazed types. Be aware that some of this tile is fired at higher temperatures than others. In general, the lighter the glaze, like the ones found in this sunny bath, the harder the finished tile.

RIGHT When selecting the tile that will help you attain your aesthetic goals, consider not only color and type, but shape. These long, narrow tiles installed horizontally lend a calming note to support the bath's Zen theme.

LEFT Typically in basic white with a uniform 3-in. by 6-in. rectangular, or a 4-in. by 4-in. square shape, subway tile is a classic, glazed ceramic tile that first gained popularity in early 1900s homes. It's back by popular demand, loved for its ability to mix with most every other tile, including this floor tile in a basket-weave pattern.

more about...
TYPES OF TILE

before you set out to shop for bath tile, familiarize yourself with the different types as identified by the Tile Council of North America.

Glazed tile is available in high-gloss, matte, and abrasive slip-resistant finishes. Hand-painted glazed tiles come in beautiful and intricate patterns, but the glaze doesn't go all the way through, so if it gets chipped, you're going to see the color inside.

Mosaic tile is made from different types of clay with color pigments added so the color goes all the way through the tile. This tile is available in glazed or unglazed finishes.

Quarry tile is made from a mixture of unglazed clays. It comes in earth tones: gray, red, and brown. The color comes from the clay as well as the temperature and duration of firing. The tile is usually porous and may stain if left unsealed.

Porcelain tile is fired at extreme temperatures, making it stronger and harder than other ceramic tiles. It is extremely wear-resistant and absorbs less water than other ceramic tile, making it an excellent choice for high-traffic areas of the home, especially those regularly exposed to moisture.

ABOVE One of the greatest advantages of tile is its design flexibility. In this shower stall, creamy porcelain tile was easily cut and trimmed to make way for an in-shower, octagonal window. Sunlight and a view to the treetops enhance the experience.

RIGHT The decorating formula is simple: For a more modern look, use large size tile. In this master bath, 12-in. and 18-in. tile create a Zen modern beauty.

LEFT Glass tile has exploded onto the design scene over the past couple of years. In a pale, bottle-glass version, the tile on this bathroom wall is a modern take on the classic subway tile.

ABOVE Tile can be cut, carved, embossed, and stamped into beautiful motifs and patterns. In this elegant space, a shell motif is the subtle between-the-floor pattern, and the border of the tiled wainscot.

LEFT Mosaic tile adds a sophisticated, yet somewhat exotic sensibility to a bathroom space. Because it is most often sold prepatterned and attached to sections of mesh, it isn't as hard to apply as it looks, even in an intricate, overlapping ring pattern such as this.

glass

Glass is one of the most popular surface materials in bathrooms—and for good reason. The material is nonporous and completely water resistant, can be tempered for hard wear, tinted, and poured into all shapes. As a design material, glass can't be beat for adding sparkle from reflected light. Clear glass allows for a through-view, preventing visual roadblocks that make your space look choppy. Easy to recycle and inert as a substance, glass is also a relatively eco-conscious option, especially if you select products made from recycled glass.

So what's the trade-off? Mineral deposits and soap residue can cling to glass, making daily upkeep necessary for maintaining shine. The good news is that there are plenty of products and tools that can help. If you have glass shower doors, a rubber squeegee will be your best friend for daily cleaning. Avoid abrasive cleaners on all glass materials, as they will scratch. For a natural cleaner, try vinegar. Or, for stubborn stains, use orange citrus cleaners.

The other important consideration is to buy quality glass, especially on shower doors, for safety reasons. And when buying glass tile, make sure you have selected a tile manufactured for its intended use. For instance, don't put a glass tile on the floor unless it is designed specifically for that use.

RIGHT Glass brightens two ways: By allowing light in and then by bouncing it around the room. A tempered-glass shower stall, such as the one in this bright master bath, will make your room appear larger than it is.

ABOVE Glass shower doors come in a variety of types, from aluminum framed to frameless. This frameless shower door is held in place with minimal hardware, which results in a cleaner, more modern look.

ABOVE Bathrooms clad in recycled glass tiles are a popular, earth-friendly choice today. This modern bath is green in more ways than one, even retaining the soft green, bottle-glass hue.

Mirrors and glass surfaces on countertops or vanities look elegant and perform decorating magic. This three-panel mirror adds dimension and plays off other reflective surfaces, such as the chrome and glass-topped vanity.

natural stone

Classic natural materials such as granite, marble, travertine, and slate are at the high-end of your surfacing options. These natural options make a strong design statement and are among the strongest and most durable materials to boot. Your toughest decision may be whether you opt for a honed or polished finish. In a nutshell, honed finishes are matte and less slippery. But the honing process will leave the stone's pores open and more susceptible to dirt build-up. Polishing makes for easier maintenance but the polish does wear off if not sealed and it's slippery when wet.

The hardest stone is granite, but that, too, needs to be sealed against stains. Marble is the hands-down favorite because there are so many looks, colors, and finishes available.

More recently, limestone and concrete have been finding their way into the hands of creative homeowners, designers, and builders. Limestone is loved for its soft-as-suede appearance, but is considered by some experts to be too porous for bath applications unless it is the compact variety. Concrete brings more flexibility to the party—it can be dyed, molded, carved, cut into tile, inlaid with fun accents, and be made to look like more expensive materials. On the down side, concrete can crack more readily than its geological counterparts.

When buying stone, be sure to buy from a reputable dealer and manufacturer. There are countless types of stone, each with a different chemical makeup.

Granite is one tough material and it has a speckled beauty that almost defines what we think of as stone. Polished to a high sheen, as in this modern bath, the stone becomes slippery when wet, so keep towels, grab bars, and bath mats handy.

FAR LEFT Travertine is technically a limestone but can be dense enough—or honed and filled to be made dense enough—to be considered a marble. As this guest bath shows, travertine has delicate veining and neutral coloration.

LEFT Honed stone is still smooth, but not as slick as its polished counterparts. The key to success with stone is making sure water can't seep behind it. Maintaining grout is as important as taking care of the stone itself.

LEFT Stone is available in a range of natural hues and a variety of shapes and sizes, from mosaic tiles to large sheets. Mixing and matching stone creates a visually appealing but still subtle patchwork of patterns and shades.

•natural stone considerations

Stone does have its personality quirks. Because stone is natural, color and veining can vary from tile to tile. Some people tire of pattern quickly and, given that this is a big investment that you'll live with for a long time, you'll want to consider your tolerance for patina and natural variations.

With the exception of some kinds of soapstone, natural stone—and any seams or grout lines—will also have to be sealed against moisture and staining. And stone is even less tolerant of non-rigid floors than ceramic is, making it more prone to cracking. Finally, consider how many butterfingers you have in the household. Due to its hardness, anything glass or delicate will undoubtedly shatter if it hits a stone floor or counter.

ABOVE Marble tile adds a footnote of elegance plus practical durability to any bath. Marble tiles of all sizes cover the floor and shower of this Iowa bath, adding visual interest and grandeur.

RIGHT Turkish limestone is an economical variety of this popular stone. Here, it is used to create a shower stall while serving as a soft, almost suede-like, backdrop to a modern tub.

ABOVE Stone tile can be as smooth as silk or as rough as burlap. The variety of honed and hand-hewn tile in this small bath adds interest, with a minimum of jarring color contrast.

LEFT Stone can adorn your bath with a profusion of pattern, or very little. In subdued and sophisticated shades of gray, this master bathroom has a clean and calming aesthetic.

vinyl

● ● ● VINYL FLOORING (ALSO KNOWN AS RESILIENT FLOORING) HAS GONE SEVERAL PACES

past "the wet look" that once prevailed in this category of coverings. The material has always been well-suited to bathroom applications due to its softness underfoot and its ability to stand up to moisture. Plus it can be installed over another layer of resilient flooring or wood when properly prepared. Many homeowners prefer its relative warmth compared to stone and tile.

Textured vinyl is also more slip-resistant and therefore a better choice when the bathroom will be used by children or the elderly. Now you don't have to sacrifice style to gain these attributes. Though still the most affordable of flooring materials, today's vinyl products come in a whole host of options in finish, texture, color, and design. You can find material that mimics the look of stone, tile, wood, or even vintage linoleum. Adaptable vinyl is also available in matte or textured finishes, and a rainbow of bright colors and artful patterns. This flooring carries a variety of price ranges, but keep in mind that the most economical grades will be the first to fade, crack, or show signs of wear.

RIGHT Vinyl flooring materials are available as sheet goods or cut into squares. This bath shows vinyl in a matte finish, which is better-suited for its quiet style.

FAR RIGHT When planning a bath for children, vinyl is an excellent choice. In this kids space, vinyl flooring will be softer underfoot and easier to scrub up than other flooring choices, such as stone or wood.

Put laminate on the vanity and vinyl on the floor and you've got an easy-clean, waterproof bathroom. Vinyl flooring acts like a chameleon, mimicking many other types of materials.

laminates

●●● LAMINATES OR SOLID-SURFACE MATERIALS make another excellent choice in the bathroom.

Laminate flooring is designed as a "floating" floor system (requiring little adhesive), made up of a fiberboard core, topped by a printed image (typically of a wood grain or type of stone), and sealed with a wear layer of clear resin.

The main advantage of laminate flooring, which generally comes in planks or tiles, is its ability to be installed over any type of floor, and it can be a dead-ringer for more expensive natural materials. One word of caution: When used in a bathroom, the areas around toilets, tubs, sinks, and showers must be sealed carefully with silicone caulk to prevent moisture from seeping under and into the unprotected edges of the flooring.

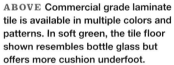

ABOVE Commercial grade laminate tile is available in multiple colors and patterns. In soft green, the tile floor shown resembles bottle glass but offers more cushion underfoot.

RIGHT For the look of marble without the high price tag, check out laminates. In a classic checkerboard pattern, this floor can easily be taken for the real thing.

FACING PAGE Laminate wood flooring offers the warmth of wood grain with easier upkeep. This laminate floor lays the groundwork for a natural style bathroom and is sealed against moisture thanks to its manufacturing process.

wood

● ● ● THOUGH MANY HOMEOWNERS SHY AWAY FROM REAL WOOD FLOORING because of its susceptibility to warping and finish wear, the superior finishing of today's woods make it a more practical option. Using the new high-tech sealants available, wood flooring can be sealed at the manufacturing stage or after it's been installed. Be aware that some hardwoods, such as teak, will stand up better to moisture than more porous woods, such as pine. Also, even with the best sealant, water should not be allowed to pool and stand on a wood floor, as it could degrade the finish over time and allow water to seep through and cause warping.

Engineered woods are also available. Rather than solid boards, engineered wood flooring is made by gluing thin layers of wood together to form sturdy, plywood-like boards. Often the top layer of the wood is a veneer of the wood of your choice. Though these products are less apt to warp, the finishes can be compromised by water and, unlike solid wood, can only be refinished once before the top veneer wears down.

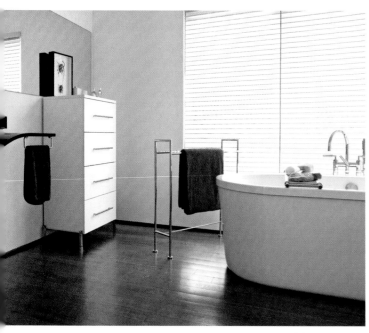

ABOVE When properly sealed against moisture, wood presents a flooring option that can take the visual chill off of the most modern of baths. In medium-tone wood grains, this polished oak floor also feels warm underfoot.

RIGHT Wood veneers are a versatile wall covering. These birch wall tiles are installed with the grains running vertically and horizontally to create an interesting but subtle checkerboard pattern.

CORK FLOORING

his familiar, tawny material is surprisingly green. Cork comes from the bark of a tree that can be harvested once every 10 years without damaging its host. It also feels good underfoot thanks to the millions of air bubbles per cubic inch waiting to cushion your soles. Cork's list of attributes goes on to include thermal and sound insulation and hypoallergenic properties.

Like its woody cousins, though, cork is susceptible to moisture, which will cause it to swell and break down. A relatively new product on the market, cork flooring also comes with a premium price tag. But if you've fallen for the rich, natural look of cork, it will be a durable choice if you protect it with industrial-grade, water-based polyurethane finish after installation to keep water out of the seams. Also, choose parquet cork tile that is glued down, rather than a floating system of planks.

TOP LEFT Certainly not just for floors, wood paneling and counter-tops have their place in the bath as well. Honey pine brings charm and a golden glow to this traditional bathroom.

ABOVE Wood is loved for the warm contrast it brings to modern bathroom fixtures. Look how beautifully oak hard-wood flooring and a clean-lined, Shaker-style vanity support a contemporary vessel sink and sleek Italian-style tub.

LEFT One of the most versatile background materials, wood can be used on walls, cabinetry, and flooring. Wood plays a starring role in this modern, but warm and natural bathroom.

wall coverings

●●● IF YOU LOVE THE ARTFUL DESIGN AND CLEAR COLORATION THAT CAN BE FOUND IN WALLPAPERS THESE DAYS, but worry that your choice might one day peel down around you due to moisture—fear not! Steam is only an issue if your bathroom's ventilation is poor, your paper is installed with gaping seams, or your walls are not properly prepared. In fact, vinyl is the material most often found in wall coverings (either solid vinyl or vinyl-coated), and we already know how water resistant it is.

To make sure you have chosen a covering that will stick around for a while, shop for varieties that can be scrubbed. More porous papers can also be used in a ventilated bathroom, though you are best advised to avoid using it directly over a shower area. To avoid condensation and the resulting mold and mildew, porous papers are actually advisable on outside walls, where temperature differences between indoors and out are the greatest. Many mildew-resistant coverings are also on the market, some of which are marked specifically for use in the bathroom.

BELOW Wall coverings can bring grand style to the smallest of baths. This silvery, tone-on-tone paper adds eye appeal without overwhelming the tiny space with pattern.

ABOVE Simple patterns in subdued hues can turn a potentially heavy, traditional room into a light and open environment. In this St. Louis bathroom, a floor-to-ceiling stripe counters the ornateness of the other furnishings.

TOP LEFT New wall coverings in vintage patterns combine old world-style with new world engineering. Easy to wipe clean, this floral toile lends prettiness and practicality to the space.

LEFT Stripes and other geometric patterns add a sense of architecture to any space. If you select a wide stripe, as in this bath, you'll create a more contemporary feeling.

ABOVE Prevent white-out in the bathroom with a light but cheery wall covering. Used on the upper wall only atop a charming wainscot, this pale blue floral adds a pretty touch of pattern without spoiling its inviting airiness.

Contrary to popular belief, you shouldn't shy away from large-patterned wall coverings in a small space. To maintain an airy attitude, as in this space, select open patterns rather than dense ones and select a soft color.

• wall coverings

Wall coverings are not only engineered to be a practical choice in today's bathrooms, they are also a design problem solver. With the right pattern, you can alter the perception of space. Select a large motif, for instance, if you want a room to appear more intimate or dramatic. To make a small space appear larger, gravitate toward smaller, more subtle prints. To make your ceiling seem higher, go with vertical lines.

Stripes and other geometric patterns make a bold statement and add a sense of architecture to a nondescript space. There are also wall coverings that add more texture than pattern to your walls, such as grass cloth, for a more interesting wall surface. No matter what type of design you select, think about your tolerance for visual busyness to determine how happy you will be with your choice in the long run. Don't know for sure? Open your closet to see how much pattern you find hanging in there.

ABOVE To add visual punch to a colorless space, go bold. This California designer mixed and matched pin-striped wallpaper with neutral tile, cabinetry, and flooring.

LEFT Wall coverings not only have a sense of style, but they can have a sense of humor as well. Chartreuse images on a pale blue background bring fresh color to this room.

LEFT Menswear patterns translated into wall coverings add a tailored crispness to a room. The oxford stripe on the ceiling also draws the eye up and makes the ceiling appear higher than it is.

painted surfaces

●●● A BRUSH AND A BUCKET OF PAINT ARE THE HOME DECORATOR'S BEST FRIENDS. In an average-sized bathroom, for less than $100 worth of materials you can pull off a dramatic makeover on a slow weekend! Choose the right kind of paint for the job and you've also dressed your walls with a coat of armor against moisture and wear.

A bathroom's high humidity can create problems for paint. When moisture condenses on walls and trim, it can cause the finish to crack and peel. High humidity can also leave uneven discoloration on paint and eventually allow mildew to develop. To avoid these and other pitfalls, choose high-quality, semi- or high-gloss varieties for your bathroom. These more nonporous paints produce smoother finishes that are easier to clean. Many manufacturers go one step further by making specially designated kitchen and bath paints. These types typically have additives that help to control mildew.

There is one drawback of high-sheen paints: The glossier the sheen, the more it will show imperfections in the underlying surface. To find a happy medium, look for flat or eggshell paints that promote easy cleaning on their labels. In general, it is better to pick latex over oil paints for interior situations because it is easier to clean and allows the walls to breathe during temperature fluctuations. Latex is also preferred when used in closed interior spaces.

Even a touch of painted-on color can make a big difference in a room. Consider painting tub alcoves, wall niches, or even the water closet in a contrasting shade to emphasize fun room features or areas within the room that perform different functions.

COLOR CONFIDENCE

Picking the type of paint you need for your bath is one thing—deciding on a paint color is quite another. For many home decorators, choosing wall colors is a stressful process. Here are a few tips that will make it easier on you:

1. To make a small bathroom appear larger and airier, pick a light paint color. Rich or dark colors make walls appear to advance for a cozier space.

2. Cosmetic shades in peach, pinks, and corals tend to flatter skin tones and make a lovely choice in bathrooms and powder rooms.

3. The direction your room faces impacts the type of light it receives. As a general rule of thumb, you can visually heat up a coolly-lit, north-facing room with warm, earth-based colors. Alternately, cool off an intensely-lit southern room with water shades of blues and greens.

4. Study the color wheel to find hues that complement each other. With a refresher course in color theory, you might discover how easy it is to choose a foolproof palette.

5. You can't go wrong selecting colors found in nature. For pure color, bring in a favorite shell, flower, stone, or whatever. Most paint shops and departments have computer color matching systems that will help you get close to the real thing.

TOP LEFT Dramatic effects can be created with paint. Given a color wash with diluted terra-cotta paint, these walls look as if they spent time under a Tuscan sun.

FAR LEFT Though painting over wood paneling gives some people pause, the results can be dramatically positive. Once a brown-on-brown attic space, this bath adopts a brighter outlook with grassy green and clean white.

LEFT Make sure your walls are smooth before you buy high-gloss paint. This gleaming red finish reflects the light in a dramatic way for an added benefit.

fabrics

● ● ● NOTHING SOFTENS THE SHARP EDGES AND COLD SURFACES OF A BATHROOM LIKE FLOWING FABRIC. Whether it's draped across the shower or covering a window, fabric is a fabulous way to link a bathroom with the design of an adjacent bedroom or simply to add more personality to the space. Color and pattern aside, the most important characteristic of a bathroom fabric is its resistance to mildew. Many fabrics are treated for stain resistance and mildew resistance at the manufacturing stage.

As with paint and wallpapers, there are fabric lines that are specifically made for kitchens and baths. Cottons are great choices in bathrooms because of their ability to be machine-washed. Also, consider the vast array of indoor/outdoor fabrics currently available. Amazing strides have been made in the design and hand of these fabrics, which also offer superior resistance to moisture, clean easily, and wear like iron.

FAR LEFT A single panel of flowing fabric might be all that's needed to soften a room of hard surfaces. Not content to be tied back in the usual way, this cinched sheer becomes the center of attention.

LEFT For softness without added pattern, there are plenty of solid fabrics to consider. A small check fabric trims this solid gray fabric with a contrasting welt and banding on the Roman shades to give the areas more definition.

FACING PAGE The right fabric choice can bring out your room's personality. Without the cabbage rose print used sparingly throughout, just think how lackluster this space would be.

window coverings

● ● ● THE NEED FOR PRIVACY IN A BATHROOM IS A NO-BRAINER. The good news is you can accomplish both privacy and prettiness at the bathroom window with the right choice in window coverings. You can also make a wimpy window look more dramatic, direct and control the light, and add energy efficiency with the right choice of window covering.

Fabric coverings are a good choice when you want to dress up a plain window or throw an angular, hard surface room a bit of curve and verve. Selecting the same or coordinating fabrics from an adjacent bedroom will establish a visual bridge to the space for a more finished look. The biggest caution with fabric coverings is to prevent the fabric from pooling on the floor where water might also be collecting.

Folds of fabric perform many functions in a bathroom. Here, draperies with lovely French pleats add privacy, lushness, and insulation. Plus, when pushed to the sides of this picture window, they frame a stunning bay view.

ABOVE Even small windows deserve some attention. This swing-out window isn't hindered by a richly-colored Roman shade in French toile and red-checked banding.

LEFT Window valances may not help with the privacy factor, but these toppers can turn a plain window into a more attractive feature. Here, the swag and jabot treatment is mounted higher than the window, at ceiling height, giving the panes more prominence.

ABOVE Plantation shutters are basically wooden blinds with a frame. These shutters swing in when the owner wants maximum sun and breeze. They also provide a link to the louvered doors, giving the room a finished look.

TOP LEFT You can inset blinds and shades within a window's frame or, if your window architecture isn't something to write home about, mask the frame completely. Mounted outside the window's edges, this fabric shade appears to be a wall of light.

ABOVE Pleated shades are available without pull strings and wands for an ultra-clean look. When fully extended, these shades are translucent, subduing the sun's rays to a pleasant glow.

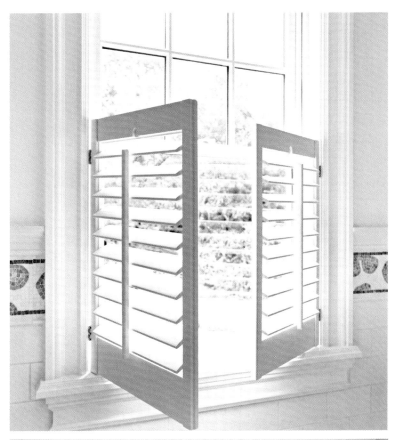

blinds and shades

For bathrooms that could benefit from more architectural character, consider wood blinds or shutters, which also offer you the best light control. Though wood treatments often will have to be custom-fabricated, once installed they are timeless in style and wear. For a similarly clean look that will also increase the energy efficiency of drafty windows, check out pleated shades.

Many pleated shades are double-layered or designed in a honeycomb pattern to add even more insulating power. Mini-blinds are another clean and classic look, but beware of inexpensive metal blinds that will rust with frequent exposure to moisture.

ABOVE When you want both privacy and light, cover only the lower half of your windows and let the sun stream in up top. Half shutters can pull off that trick, or swing fully open on piano hinges.

LEFT Custom shutters are a good choice for unusual windows. These plantation shutters accentuate the curves of this alcove's unique vintage windows.

lighting, heating, and ventilation

7

• ○ ○

THOUGH LIGHT BULBS, RADIATORS, AND EXHAUST FANS MAY NOT BE THE most exciting or beautiful elements of a luxurious bath, they certainly are key to the comfort that a new spa-like space is ultimately all about. These nitty-gritty details, after all, might seem more important if you thought of them as soft mood lighting, warm towels, and a cozy floor underfoot.

Lighting, for example, involves not only switches and shades, but windows as well. Like every other room in the house, windows and natural light are essential design elements that should be discussed when drawing up floor plans. Is there a view? If so, will there be privacy? What about transoms or skylights? At night there is much to consider, too. Task lighting near mirrors, overhead fixtures, dimmers, and sconces all need to be thought out with both aesthetics and practicality in mind.

And heating and ventilation are just as important. You don't want to step out of a hot tub or shower only to stand and shiver or feel as if you're dripping in a steam bath instead of your own home. So before everything else is picked out and put into your plans, spend time thinking through these three key components. In the end, they will make your new bath feel as good as it looks.

Multiple sources of light and heat in a bathroom mean multiple sources of comfort. Natural and artificial light create a sunny attitude here. A radiant-heat floor and a towel warmer carry through that warm feeling.

lighting options

● ● ● IF POSSIBLE, PLANNING FOR YOUR NEW BATH SHOULD INCLUDE OPTIONS FOR NATURAL AND ARTIFICIAL LIGHT. With windows or skylights, first determine if you want them to be operable. Though fresh air is wonderful, depending on your location, outside air quality (and possibly noise and safety) may be a concern. Next, choose a style suited to your house and determine size and placement. The most beautiful windows in the world will look awkward if they are out of proportion or oddly positioned. For artificial lighting, there are myriad factors to take into account such as style of fixtures, placement, function, timers, dimmers, and bulbs.

ABOVE Light has the power to transform a dull bath into a dramatic one. By designing a vanity niche that isn't quite flush with the wall and adding simple strip lighting in the recessed area, these homeowners created unique style.

ABOVE Uplights, downlights, all around lights, and skylights—all types add beauty and function to a bath space. In this open bath design, lighting helps define the activities within the space as well.

GREEN OPTIONS

O ne way to be more eco-friendly— both here and throughout the house—is by incorporating energy-efficient lighting and dimmer switches. This simple change will not only save energy, but extend the life of a traditional light bulb as well.

Compact fluorescent bulbs, timers, and motion-detecting switches are other items that cut energy use, too. On-demand water heaters are another great energy-efficient item for the bath. Though initially more expensive to install than a traditional tank heater, long-term use will pay back the higher investment cost with reduced energy use.

ABOVE "Can" lights direct bright beams and extra warming power right to where it's wanted. When placing lights within or near the shower, make certain the fixtures you choose resist rust damage, as these stainless models will.

RIGHT Specialty rods combined with dramatic lighting can transform your space into a one-of-a-kind experience. This circular rod and recessed tinted lighting, plus the unique window-scaping, shows how to use light for an enhanced effect.

BELOW Natural and overhead light are examples of ambient lighting. This San Francisco home uses tempered-glass interior walls to make a modern design statement and to escort natural light in throughout the space.

• ambient light

In the bathroom, good lighting is crucial. Shaving and putting on makeup are two practical examples that come into play every day. But a moonlit or sunset soak in the tub can be just as essential to your routine. For that languid end-of-the-evening soak, windows are needed, of course. Plus, who wouldn't agree that one can never have enough natural light and fresh air?

Architecture should dictate your style selection. For example, divided-light, double-hung windows make an ideal match for Craftsman-style homes, and large, fixed, single panes fit right in with contemporary homes. Placement will determine your privacy needs. Think window treatments or opaque glass as solutions that will help you keep windows in your plans if privacy is an issue.

TOP LEFT Skylights and overhead lighting combine to heat and light up a bathroom by day or by night. This square skylight is inoperable and well sealed to avoid letting in moisture where it isn't wanted.

ABOVE To gain privacy and sunlight in your bath area, consider mounting window dressings to three-quarter height. This tub alcove has it all—a view of the sky, rays of warming sunshine, and a wall-mounted, flat-screen television for another layer of enjoyment.

LEFT Creative solutions let you have your windows and use them, too. A mirror cleverly installed in front of a sliding casement window allows natural light to fill the space and work as sidelights for additional task lighting.

Building out from the wall will allow you to create dramatic and useful lighting effects. In this well-lit master bathroom, glittering sconces enhance the romantic look of the space and light a recessed area for focal-point drama.

• task lighting

When planning the lighting for your bath, task lighting should be of first importance. This type of lighting nearly always translates into the vanity lights. There's nothing more frustrating than trying to put on makeup (or shaving, for men) in front of a poorly-lit mirror. As a rule, sconces should flank a mirror at eye level about 36 in. apart to prevent distracting shadows on the face; lights above the mirror or over the sink will not illuminate the face properly.

Depending on the floor plan and the size of the room, task lighting may also be needed in a toilet area or a shower stall (be sure the fixtures are designed for use in humid areas). Overhead, decorative, and accent lighting are needed as well, but driven more by look than function. Pendants, cove lights, chandeliers, and spots are all items that can enhance the design and brightness of a room.

for old-house lovers, renovation work presents an opportunity to update the mechanical workings of heating, plumbing, and electrical systems. Sometimes, though, local contractors or DIYers may not be attuned to all the beautiful reproduction hardware available to finish out a new space in an appropriate or complementary style.

For period-look switches and switch plates, light fixtures and shades, vent grilles, and more, try scouting the Web for products that suit your architecture. Or contact a local antique lighting specialist to find the real thing that has been refurbished to modern electrical code.

ABOVE Task lighting will live up to its name only if installed properly. Though many vanities are lit from above, this bathroom shows the best placement—to either side of the mirror.

TOP Recessed or track lighting can also be directed to serve as task lighting. These rotating spotlights can be adjusted to wash a wall with light, or to be a backup for the simple sconces.

ABOVE The best task lighting performs a design function and directs light where needed most. Symmetry is a virtue in this warm, contemporary room. Note how simple cylinders of light reinforce that balanced feeling.

• more lighting options

RIGHT The most effective lighting arrangements use a variety of light sources and in surprising ways. The cool white scheme in this Palm Springs bath is balanced by the warm glow from vintage-modern pendant lights and a bright green view through a clear and clever placement of a sliding shower door to the outdoors.

BELOW Modern, high-end tubs are taking the bathing experience to the next level with colored accent lights designed to soothe. This custom granite tub is accented from within and outside of the tub, making the lighting a centerpiece in this room's design.

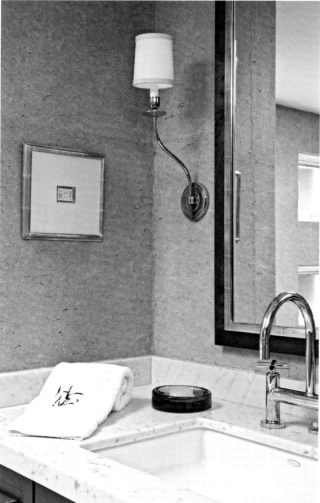

TOP LEFT Reflective surfaces can also be considered a light source—and there are certainly many of these in a bath scheme. Mirrored tiles in this sparkly bath virtually double the light of the room.

LEFT Consider lighting as the sparkling jewelry of a bathroom scheme and coordinate it to the style and scale of your space. Reaching out from the wall to put light where it's most effective, this delicate sconce adds pretty design and practical lighting in a sliver of space.

ABOVE When the three types of light—task, ambient, and natural—come together, space is literally transformed. This small bath might have appeared flat and colorless, but it takes on a golden glow with only a simple lighting plan of recessed light, vanity sconces, and natural light through an unfettered window.

RIGHT Cosmetic mirrors add a pretty detail and practical magnification. Wall mounted and hinged, this classically styled mirror swings out when in use, or tucks against the wall and out of the way when its services are not needed.

BELOW To soften task lighting, put it on a dimmer or purchase lighting with shades. Here, the owner opted for both to apply cosmetics correctly for daytime or evening outings.

MIRROR OPTIONS

mirrors have long been a standard component in a decorator's bag of tricks, performing visual magic in the bathroom. Mirrors grab the light and bounce it around the room, increasing the bounty of artificial and natural light. A sizable looking glass can also be used to make a cramped bath appear to double in size by reflecting the room back at all who look into it. A series of contemporary mirrors or one large, ornate mirror also adds dimension and architecture to a bland space with little effort.

There are a number of ways to install bathroom mirrors. Your local glass shop can cut a mirror in almost any size or shape that can be mounted flush against a wall. Consider hiring a professional to drill through the material using special tools that will allow you to then mount lighting right over the mirrored glass for a lovely effect. Or you might like the look of a framed mirror that is either fixed or pivots.

ABOVE When you want a wall-sized mirror and the right task lighting, there is an option. These sconces have been mounted directly to the mirror, virtually doubling the light.

TOP RIGHT A mirror that is framed and beveled adds a touch of class *and* a touch of glass to a space. In this case, the quality mirror hides a handy medicine chest that's been inset between the studs of the wall.

• heat

When it comes to heat in the bathroom, simple experience tells us that it's more comfortable to have the room slightly warmer than the temperature in the rest of the house. There are a number of ways to achieve this, but a growing trend in bath design is the installation of radiant floor heating. If that's outside of your budget or simply not feasible, consider a simple wall-mount electric heater or a heat lamp.

If the sky's the limit—and you have enough square footage—design the space to include a gas fireplace. For a little luxury, consider a towel warmer. There are plug-in types or more elaborate systems that can be wired into your room's electrical system.

ABOVE Turn up the heat in a house with a closet-sized sauna. This one stands next door to a steam shower for the ultimate in warmth.

RIGHT The benefits of adding a towel warmer are pretty apparent. Just recall those special times when Mom pulled a towel right out of the dryer and tossed it over your shoulders. This wall-mounted system adds a pretty design feature, too.

more about...
RADIANT FLOOR HEATING

to get that pampered feeling in the bath, there's nothing better than stepping out of the tub or shower, especially during the winter months, onto a cozy, heated floor. If you're planning a bath renovation or even building a whole new house, radiant floor heat is a great way to get that sensation. You may want to consider an electric system rather than hydronic radiant floor heat if up-front cost is a concern and you plan to treat your feet in just a room or two.

Why? Hydronic or liquid systems need yards and yards of tubing, boilers, pumps, and more, rendering them expensive to install, repair, and maintain. Electric systems, especially for smaller-scale baths, are generally easier to install and less expensive than hydronic systems. In general, an electric system will cost under $1,000 to install, compared to several thousand dollars for a comparable hydronic floor.

A steam shower obviously contributes heat to a space. When trapping steam, it's imperative that all surfaces in the room be sealed. Though the wood paneling on the barrel-vault ceiling and elsewhere in the room might seem a risky choice next to this stone steamer, it's as polished as a yacht's deck and will ward off moisture damage.

When privacy isn't an issue, why not grab the most light and air you can? Facing a wall of green vines, this tub area needed light and ventilation more than privacy, so the owners opted for a wall of French doors.

• humidity

A key part of any bathroom renovation should be the installation of a properly powered exhaust fan. Without one, humid air from hot baths and steamy showers will invariably lead to mold or mildew—and eventually rot if the room is chronically damp. Ideally, you should install a fan that has a feature which engages the fan automatically at a preset high humidity level. Using fans with timer switches is another good way to ensure that the bath is free of moisture. Timers also prevent the needless

waste of energy that occurs when the fan is left on by mistake.

When researching fans, know that they are rated according to the volume of air they can move per minute (known as CFM, or cubic feet per minute). Rough guideline: A small bath should have a fan rated at 50 CFM; a larger bath over 100 sq. ft. will need at least 100 CFM or more. When buying a fan, pay attention to noise, too. Fan sound is measured in sones; look for one rated 1 sone or less.

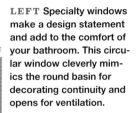

BELOW One way to increase circulation and add to your home's design flow is to use partial walls that will allow the air to move throughout a larger space. Here, a vanity and mirror appear to float in the center of the room, creating a divider without adding the visual blockade a full wall would.

LEFT Specialty windows make a design statement and add to the comfort of your bathroom. This circular window cleverly mimics the round basin for decorating continuity and opens for ventilation.

ABOVE A window in the shower, when properly sealed against moisture, has a couple of benefits. Here, a large sliding window invites in fresh air and ushers out humidity.

climate control

● ● ● OF ALL THE ROOMS IN A HOUSE, the bath is the one that needs to be absolutely comfortable in terms of temperature and humidity. Quite frankly, that's because it's most likely the one room in which you spend a good deal of time undressed. Exposed skin is sensitive to water temperature, the warmth or cold of the floor, the moisture in the air, and even the temperature of the toilet seat.

Such basic sensations then lead to the decisions you need to make in the planning process of any renovation or building project. Fans, floor or supplemental heating systems, and water fixtures with thermostatic valves are all items that should be on your punch list. Many of these items, happily, don't come with a stiff price tag.

ABOVE In nature, stone surfaces absorb heat from the sun by day, then release it back into the environment long after the sun goes down. Sunny southern-facing windows mean this trick works inside, too.

RIGHT Before you give old radiators the heave-ho, consider how good that warm, moist heat would feel in the bathroom. This clever homeowner decided both radiant and dry heat were the way to go.

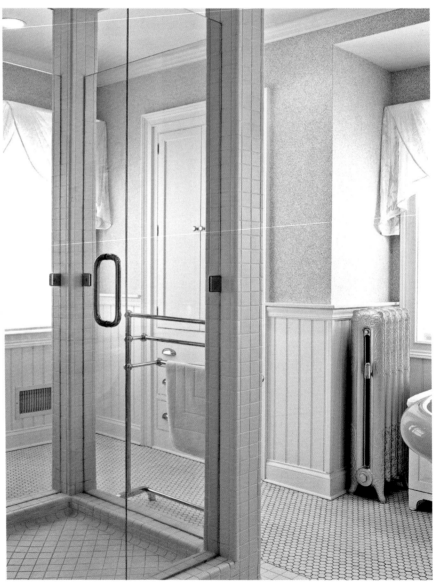

Find creative ways to trap and hold the heat from a hot shower. Here, two floor screens act like a big hug, surrounding this clawfoot tub—and the person in it—with cozy warmth and privacy.

resources

All Granite and Marble Corporation
1A Mt. Vernon St.
Ridgefield Park, CT 07660
201-440-6779
www.marble.com

American Standard
Customer Care
1 Centennial Plaza P.O. Box 6820
Piscataway, NJ 08855-6820
1-800-442-1902
www.americanstandard-us.com

Ann Sacks Tile
800-278-8453
www.annsacks.com

Armstrong World Industries
2500 Columbia Ave.
P.O. Box 3001
Lancaster, PA 17604
717-397-0611
www.armstrong.com

Ask the Builder
P.O. Box 887
Meredith, NH 03253
www.askthebuilder.com

Build Direct
877-631-2845
www.builddirect.com

Consumer Reports
101 Truman Ave.
Yonkers, NY 10703-1057
866-208-9427
www.ConsumerReports.org

Do It Yourself.com
909 North Sepulveda Blvd.
11th Floor
El Segundo, CA 90245
800-692-2200
www.doityourself.com

Eljer, Inc.
4801 Quorum Dr.
Dallas, TX 75254
972-560-2000 or 800-423-5537
www.eljer.com

Grohe America, Inc.
241 Covington Dr.
Bloomingdale, IL 60108
630-582-7711
www.groheamerica.com

The Home Depot
800-553-3199
www.homedepot.com

HomeTips
3715 Market St., Ste. 205
Glendale, CA, 91208
www.hometips.com

Jado/Porcher
Customer Care
6615 West Boston St.
Chandler, AZ 85226
800-359-3261
www.jadoporcher.com

Kohler Co.
444 Highland Dr.
Kohler, WI 53044
800-456-4537 or 920-457-4441
www.kohler.com

Lowes
Customer Care
P.O. Box 1111
North Wilkesboro, NC 28656
800-445-6937
www.lowes.com

Marble Institute of America
28901 Clemens Rd., Ste. 100
Cleveland, OH 44145
www.marble-institute.com

National Association of Home Builders
1201 15th St. NW
Washington, DC 20005
800-368-5242
www.nahb.org

**National Association of the
Remodeling Industry (NARI)**
780 Lee St., Ste. 200
Des Plaines, IL 60016
800-611-6274
www.nari.org

National Kitchen and Bath Association
687 Willow Grove St.
Hackettstown, NJ 07840
800-843-6522
www.nkba.org

National Wood Flooring Association
111 Chesterfield Industrial Blvd.
Chesterfield, MO 63005
www.woodfloors.org

Plumbing-Heating-Cooling Association
180 S. Washington St.
P.O. Box 6808
Falls Church, VA 22046
800-533-7694
www.phccweb.org

ShowerSart LLC
Evolve Showerheads
16309 N 82nd St.
Scottsdale, AZ 85260
480-496-2294
www.evolveshowerheads.com

U.S. Green Building Council
1800 Massachusetts Ave., NW
Suite 300
Washington, DC 20036
800-795-1747
www.greenhomeguide.com

Wilsonart International, Inc.
2400 Wilson Pl.
P.O. Box 6110
Temple, TX0 76503-6110
800-433-3222
www.wilsonart.com

photo credits

pp. ii–iii: Photo © Eric Roth

p. iv: (top left) Photo © Eric Roth; (top right) Photo © Mark Lohman; (bottom left) Dan Duchars/Redcover.com; (bottom right) Jean Maurice/Redcover.com

p. v: (top left) Photo © Eric Roth, Design: www.decoridesigns.com; (top right) Photo © Eric Roth, Design: www.trikeenan.com; (bottom left) Photo © Mark Samu, Design: Mosaik Design

p. 2: (left) Photo © Tria Giovan; (right) Photo © David Duncan Livingston

p. 3: Sarah Hogan/Redcover.com

CHAPTER 1

p. 4: Photo © davidduncanlivingston.com

p. 5: (top to bottom) Photo © Olson Photographic, Design: Brindisi & Yaroscak, Darien, CT; Photo © David Duncan Livingston; Photo © Tria Giovan; Photo © Mark Lohman

p. 6: Photo © Eric Roth, Design: www.fbnconstruction.net

p. 7: (top left) Photo © Holly Joliffe/Redcover.com; (top middle) Photo © Eric Roth; (top right) Photo © Eric Roth, Design: http://www.benjaminnutter.com/; (bottom) Photo © Nina Assam/Redcover.com

p. 8: (left) Photo © Winfried Heinze/Redcover.com; (right) Photo © Douglas E. Smith

p 9: Photo © Tria Giovan

p. 10 Photo © Olson Photographic, Design: P&H Construction, Norwalk, CT

p. 11: (left) Photo © Jean Maurice/Redcover.com; (right) Photo © Douglas E. Smith

p. 12: Photo © Sarah Hogan/Redcover.com

p. 13: (top left) Photo © David Duncan Livingston; (right) Photo © Eric Roth, Design: www.susansargent.com; (bottom left): Photo © Olson Photographic, Design: Joe Cugno Architects, Wilton, CT

p. 14: Photo © Olson Photographic, Design: Brindisi & Yaroscak, Darien, CT

p. 15: (top) Photo © Olson Photographic, Design: Heartstone Homes, South Glastonbury, CT; (bottom) Photo © King Au

p. 16: Photos © David Duncan Livingston

p. 17: (left) Photo © Eric Roth; (right top) Photo © Jon Jensen; (right bottom) Photo © Rob Karosis

p. 18: (top) Photo © Alise O'Brien; (bottom) Photo © Eric Roth

p. 19: (top) Photo © Olson Photographic, Design: Amazing Spaces, Briarcliff Manor, NY Design; (bottom) Photo © Tria Giovan

p. 20: (top left) Photo © Mark Samu; (top right) Photo © Alise O'Brien; (bottom left) Photo © David Duncan Livingston

p. 21: Photo © Mark Lohman

p. 22: Photo © Tria Giovan

p. 23: (top left) Photo © Eric Roth, Design: Horst Buchanon- www.horstbuchanon.com; (top right) Photo © Tria Giovan; (bottom) Bieke Claessens/Redcover.com

p. 24: Photo © Eric Roth

p. 25: (top) Photo © Eric Roth, Design: www.susansargent.com; (bottom left & right) Photo © Mark Samu

p. 26: (left) Photo © Eric Roth, Design: www.susansargent.com; (right) Photo © Eric Roth

p. 27: (left) Photo © Eric Roth, Design: www.susansargent.com; (right) Photo © Eric Roth, Design: www.poore-co.com

p. 28: (left) Winfried Heinze/Redcover.com; (right) Photo © Tria Giovan, Design: Suzanne Kesler

p. 29: (top) Warren Smith/Redcover.com, Design: FLOW flowlondon.co.uk; (bottom) Photo © Tria Giovan

p. 30: (left) Photo © Olson Photographic, Design: Paul Shainberg Architects, Greenwich, CT; (top right) Photo © Eric Roth; (bottom right) Photo © Olson Photographic, Design: Olga Adler Interiors, Ridgefield, CT

p. 31: Photos © Eric Roth

CHAPTER 2

p. 32: Photo © Douglas E. Smith

p. 33: (top to bottom) Photo © Douglas E. Smith; Bieke Claessens/Redcover.com; Photo © David Duncan Livingston; Photo © Olson Photographic, Design: Sally Scott Interiors, Guilford, CT

p. 35: Photo © Olson Photographic, Design: Lisa Newman Interiors, Providence, RI

p. 38: (left) Photo © David Duncan Livingston; (right) Trine Thorsen/Redcover.com

p. 39: Bieke Claessens/Redcover.com

p. 40: Bieke Claessens/Redcover.com

p. 41: (left) Photo © David Duncan Livingston; (right) Courtesy of Kohler Co.

p. 42: Photos © David Duncan Livingston

p. 43: © Olson Photographic, Design: Sally Scott Interiors, Guilford, CT

p. 44: Photo © Douglas E. Smith

p. 45: (top) Photo © Douglas E. Smith; (bottom) Photo ©King Au

p. 46: Photo © Olson Photographic, Design: Nina Cuccio-Peck Architects, Old Lyme, CT

p. 47: (top left) Photo © Todd Caverly; (top right) Photo © Olson Photographic, Design: Kitchen & Bath Designs by Betsy House, Stonington, CT; (bottom left) Photo © Ken Gutmaker; (bottom right) Photo ©

Mark Lohman, Design: Harte Brownlee and Associates

p. 48: (left) Photo © Eric Roth; (right) Christopher Drake/Redcover.com

p. 49: (top) Bieke Claessens/Redcover.com; (bottom) Photo © Eric Roth

pp. 50–51: Photos ©King Au

p. 52: Photo © Eric Roth

p. 53: (top) Jake Fitzjones/Redcover.com; (bottom) Photo © Olson Photographic, Design: Sharon McCormick Interiors, Durham, CT

p. 54: (top) Photo © Tria Giovan; (bottom) Photo © Douglas E. Smith

p. 53: Photo © Tria Giovan

CHAPTER 3

p. 56: Bieke Claessens/Redcover.com

p. 57: (top to bottom) Simon Scarboro/Redcover.com; Photo © David Duncan Livingston; Photo © David Duncan Livingston; Photo © Eric Roth, Design: www.benjaminnutter.com

p. 58: Jean Maurice/Redcover.com

p. 59: Photo © Tria Giovan

p. 60: Photo © Mark Lohman

p. 61: (left) Sarah Hogan/Redcover.com; (top right) Dan Duchars/Redcover.com; (bottom right) Bieke Claessens/Redcover.com

p. 62: Tria Giovan/redcover.com, Design: Michael Landrum

p. 63: (top left) Ken Hayden/Redcover.com, Design: Lori Dennis of Dennis Design Group; (top right) Winfried Heinze/redcover.com, Design: Marion Lichtig; (bottom left) Alun Callender/Redcover.com, Stylist: Melanie Molesworth; (bottom right) Ken Hayden/Redcover.com, Design: Lori Dennis of Dennis Design Group

p. 64: (left) Photo © Eric Roth, Design: www.benjaminnutter.com; (right) Photo © Alise O'Brien

p. 65: (left) Photo © David Duncan Livingston; (right) Photo © Eric Roth, Design: Heidi Pribell - www.heidipribell.com

p. 66: (left) Photo © Eric Roth, Design: Dressing Rooms; (right) Photo © Mark Lohman

p. 67: Photo © Eric Roth, Design: www.decoridesigns.com

p. 68: (left) Photos © Olson Photographic, Design: Studio DiBerardino, Darien, CT; (right) Photo © Eric Roth

p. 69: Photo © Eric Roth, Design: www.heatherwells.com

p. 70: (left) Photo © Eric Roth; (right) Photo © Tria Giovan

p. 71: Photos © Tria Giovan

p. 72: Photo © Eric Roth, Design: www. lesliefineinteriors.com

p. 73: (top) Nina Assam/Redcover.com; (bottom left) Photo © David Duncan Livingston; (bottom right) Photo © Tria Giovan

p. 74: Winfried Heinze/Redcover.com

p. 75: Photos © Tria Giovan

p. 76: (left) Michael Freeman/Redcover.com; (top right) Anthony Parkinson/Redcover.com; (bottom right) Sasfi Hope-Ross/Redcover.com

p. 77: (top) Paul Ryan-Goff/Redcover.com, Design: Kastrup & Sjunnesson; (bottom) Warren Smith/Redcover.com

p. 78: (left) Bieke Claessens/Redcover.com; (right) Photo © David Duncan Livingston

p. 79: Photo © David Duncan Livingston

p. 80: (left) Dan Duchars/Redcover.com; (right) Bieke Claessens/Redcover.com

p. 81: (top) Photo © Eric Roth; (bottom) Bieke Claessens/Redcover.com

p. 82: Photo © Alise O'Brien

p. 83: (top) Photo © David Duncan Livingston; (bottom) Photo © Tria Giovan

p. 84: Photo © David Duncan Livingston

p. 85: (top & bottom left) Photo © David Duncan Livingston; (bottom right) Photo © Eric Roth

p. 86: Simon Scarboro/Redcover.com

p. 87: (left) Photo © Tria Giovan; (top right) Alun Callender/Redcover.com; (bottom right) Photo © Eric Roth

p. 88: (left) Photo © Douglas E. Smith; (right) Photo © King Au

p. 89: Photo © Eric Roth

CHAPTER 4

p. 90: Photo © David Duncan Livingston

p. 91: (top to bottom) Paul Ryan/Redcover. com; Paul Ryan/Redcover.com; Photo © Jon Jensen, Design: Mosaik Design; Photo © Mark Samu

p. 92: (left) Photo © Mark Lohman; (right) Johnny Bouchier/Redcover.com

p. 93: (top) Photo © Jon Jensen; (bottom) Photo © Eric Roth, Design: www.catalanoinc. com

p. 94: Troy Campbell/Redcover.com

p. 95: (top left) Andrew Wood/Redcover.com; (top right) Photo © King Au; (bottom left) Tria Giovan/Redcover.com; (bottom right) Photo © David Duncan Livingston

p. 96: Courtesy of Kohler Co.

p. 97: (top) Photo © Jon Jensen, Design: Mosaik Design; (bottom) Courtesy of American Standard

p. 98: Photos © David Duncan Livingston;

p. 99: (left) Photo © Mark Samu; (top & bottom

right) Paul Ryan/Redcover.com

p. 100: Photo © Alise O'Brien

p. 101: (top left) Dan Duchars/Redcover.com; (top right) Photo © Mark Samu; (bottom left) Debi Treloar/Redcover.com; (bottom right) Photo © Douglas E. Smith

p. 102: Photos © David Duncan Livingston

p. 103: Photo © David Duncan Livingston

p. 104: (left) Photo © Eric Roth; (right) Photo © David Duncan Livingston

p. 105: Coutesty of Kohler Co.

p. 106: (left) Photo © Eric Roth; (center) Photo © David Duncan Livingston; (right) Coutesty of Kohler Co.

p. 107: Photo © Mark Samu

p. 108: (left) Courtesy of Kohler Co.; (right) Photo © Olson Photographic, Design: Sally Scott Interiors, Guilford, CT

p. 109: (top) Coutesty of Kohler Co.; (bottom) Photo © Mark Samu

p. 110: (left) Photo © Mark Lohman, Design: Kathryne Designs Interior Design; (right) Photo © Olson Photographic, Design: Nanci Paige Designs, Pacific Palisades, CA

p. 111: (left) Grey Crawford/Redcover.com; (right) Photo © Jon Jensen

p. 112: (left) Jumping Rocks/Redcover.com (right) Photo © Douglas E. Smith

p. 113: (left) Photo © Tria Giovan; (top right) Photo © David Duncan Livingston; (bottom right) Photo © Douglas E. Smith

p. 114: Photo © Olson Photographic, Design: Putnam Kitchens, Greenwich, CT

p. 115: (top left) Coutesty of Kohler Co.; (bottom left) Photo © David Duncan Livingston; (bottom right) Coutesty of GROHE

p. 116: (left) Photo © David Duncan Livingston; (right) Paul Ryan/Redcover.com

p. 117: (top) Paul Ryan/Redcover.com; (bottom) Photo © Eric Roth

p. 118: Photos © David Duncan Livingston;

p. 119: (top left) Photo © Alise O'Brien; (top right) Jumping Rocks/Redcover.com; (bottom) Photo © Alise O'Brien

p. 120: (left) Niall McDiarmid/Redcover.com; (right) Photo © Eric Roth, Design: Dressing Rooms

p. 121: (left) Photo © Mark Samu; (right) Photo © Douglas E. Smith

p. 122: (left) Verity Welstead/Redcover.com; (right) Photo © Mark Samu

p. 123: (top) Bieke Claessens/Redcover.com; (bottom) Andrew Twort/Redcover.com

p. 124: (left) Evan Sklar/Redcover.com; (right) Photo © Eric Roth, Design: Tricia McDonagh Interior Design

p. 125: (top) Paul Ryan/Redcover.com; (bottom) Photo © Mark Lohman

p. 126: Photo © Alise O'Brien

p. 127: (top) Photo © King Au; (middle)

Courtesy of Kohler Co.; (bottom) Courtesy of American Standard

p. 128: (left) Photo © David Duncan Livingston; (right) Courtesy of Kohler Co.

p. 129: Photo © David Duncan Livingston

p. 130: Bieke Claessens/Redcover.com

p. 131: (left) Alun Callender/Redcover.com; (right) Courtesy of Kohler Co.

CHAPTER 5

p. 132: Photo © Eric Roth

p. 133: (top to bottom) Photo © Eric Roth; Photo © Mark Lohman; Photo © Eric Roth; Photo © Olson Photographic

p. 134: (left) Photo © Eric Roth; (right) Photo © David Duncan Livingston

p. 135: (left) Photo © Olson Photographic; (top right) Photo © David Duncan Livingston; (bottom right) Photo © Eric Roth, Design: Tricia McDonagh Interior Design

p. 136: Photo © Jon Jensen

p. 137: (top) Photo © Todd Caverly, Design: G.M. Wild Construction Inc.; (bottom) Photo © Eric Roth, Design: www.kmarshalldesign.com

p. 138: Photo © Olson Photographic

p. 139: (top left) Photo © Mark Lohman; (bottom left) Redcover.com; (top right) Photo © Douglas E. Smith; (bottom right) Photo © Douglas E. Smith, Design: Harte Brownlee & Associates Interior Design

p. 140: (top left) Photo © Olson Photographic, Design: Amazing Spaces, Briarcliff Manor, NY; (top right) Sarah Hogan/Redcover.com; (bottom) Grey Crawford/Redcover.com

p. 141:(top left) Photo © Eric Roth, Design: www.trikeenan.com; (bottom left) Photo © Mali Azima; (right) Photo © David Duncan Livingston

p. 142: Photo © Mark Lohman

p. 143: (left) Photo © Eric Roth, Design: Joan Krainin Interior Design; (right) Photo © Alise O'Brien

p. 144: (left) Photo © Mark Samu, Design: Sally Scott Interiors, Guilford, CT; (right) Photo © Eric Roth, Design: www.frankroop.com

p. 145: (top) Photo © Douglas E. Smith; (middle from left to right) Photo © Douglas E. Smith; (bottom left) Photo © Jon Jensen; (bottom right) Photo © Olson Photographic, Design: Jack Rosen Custom Kitchens/Joe Currie Designs, Rockville, MD

p. 146: Photo © David Duncan Livingston, Design: Mosaik Design

p. 147: (top left) Photo © Eric Roth; (bottom left) Photo © Eric Roth; (right) Photo © Tria Giovan

p. 148: Jean Maurice/Redcover.com

p. 149: (top left) Photo © Mark Lohman; (bottom left) Alun Callender/Redcover.com; (right) Photo © David Duncan Livingston

pp. 150–151: Photos © Douglas E. Smith

p. 152: (left) Photo © Mark Samu; (right) Photo © Olson Photographic, Design: Kitchen & Bath Design Consultants, West Hartford, CT

p. 153: Photo © Olson Photographic, Design: Olga Adler Interiors, Ridgefield, CT

p. 154: (top) Photo © Douglas E. Smith; (bottom left) Photo © Tria Giovan; (bottom right) Photo © King Au

p. 155: (top) Photo © Douglas E. Smith; (bottom) Photo © Tria Giovan

p. 156: (top left) Photo © Jon Jensen, Design: Mosaik Design; (bottom left) Photo © Mark Lohman; (right) Photo © Eric Roth

p. 157: Photo © David Duncan Livingston

p. 158: (top) Photo © Eric Roth; (bottom) Photo © Mark Samu

p. 159: (top left & right) Photo © Mark Samu; (bottom left) Alun Callender/Redcover.com; (bottom right) Photo © Mark Samu

CHAPTER 6

p. 160: Photo © Olson Photographic, Design: www.ModArchitecture.com

p. 161: (top to bottom) Photo © Eric Roth, Design: www.chriswalsharchitect.com; Photo © Jon Jensen; Courtesy of Wilsonart; Courtesy of Thibault

p. 162: (left) Photo © Mark Lohman; (right) Photo © Eric Roth, Design: www.trikeenan.com

p. 163: (left) Photo © Alise O'Brien; (right) Photo © Eric Roth, Design: Heidi Pribell - www.heidipribell.com

p. 164: Photo © Eric Roth, Design: www.trikeenan.com

p. 165: (left) Photo © Eric Roth, Design: Heidi Pribell - www.heidipribell.com; (top right) Photo © Eric Roth, Design: www.chriswalsharchitect.com; (bottom right) Photo © Eric Roth, Design: Heidi Pribell - www.heidipribell.com

p. 166: (left) Photo © Eric Roth, Design: www.designlabarch.com (right) Photo © Eric Roth

p. 167: (top) Photo © Eric Roth; (bottom) Photo © Eric Roth, Design: www.trikeenan.com

p. 168: (left) Photo © Douglas E. Smith; (right) Ed Reeve/Redcover.com

p. 169: (left) Courtesy of Ann Sacks; (top right) Scott Van Dyke/Redcover.com; (bottom right) Photo © Eric Roth, Design: www.lesliefineinteriors.com

p. 170: (top left & right) Photo © David Duncan Livingston; (bottom) Photo © Olson Photographic, Design: Total Design Source, Old Saybrook, CT

p. 171: Photo © Tria Giovan, Design: Diamond Baratta

p. 172: Photo © Eric Roth, Design: Heidi Pribell - www.heidipribell.com

p. 173: (top left) Photo © Olson Photographic, Design: CK Architects, Guilford, CT; (top right) Photo © David Duncan Livingston; (bottom)

Photo © Tria Giovan

p. 174: (left) Photo © Douglas E. Smith; (right) Photo © King Au

p. 175: (left) Photo © Olson Photographic, Design: Sharon Cameron Lawn Interiors, Newport, RI; (right) Photo © Eric Roth, Design: www.trikeenan.com

pp. 176–177: Photos © David Duncan Livingston

p. 178: Courtesy of Wilsonart

p. 179: (left) Photo © Eric Roth; (right) Dan Duchars/Redcover.com

p. 180: Photos © Eric Roth

p. 181: (top) Photo © Mark Samu, Design: Paul Shainberg Architects, Greenwich, CT; (bottom left) Photo © Eric Roth, Design: www. trikeenan.com; (bottom right) Photo © David Duncan Livingston

p. 182: (left) Photo © Eric Roth, Design: www.susansargent.com (right) Photo © Alise O'Brien

p. 183: (top left) Photo © Olson Photographic, Design: The Tile Shop, Ridgefield, CT; (bottom left) Photo © Eric Roth; (right) Photo © Olson Photographic, Design: Paul Shainberg Architects, Greenwich, CT

p. 184: Courtesy of Thibault

p. 185: (top & bottom left) Photos © Mark Lohman, Design: Barclay Butera Inc.; (bottom right) Courtesy of Thibault

p. 186: Warren Smith/Redcover.com

p. 187: (top) Photo © Eric Roth; (bottom left) Photo © Eric Roth, Design: Salt Water Design Group; (bottom right) Photo © Mark Lohman

p. 188: (left) Alun Callender/Redcover.com; (right) Photo © Olson Photographic, Design: Ladd Design, New Canaan, CT

p. 189: Alun Callender/Redcover.com

p. 190: Photo © David Duncan Livingston

p. 191: (left) Photo © David Duncan Livingston; (right) Photo © Mark Lohman

p. 192: (top left) Bieke Claessens/Redcover.com; (bottom left) Photo © Jon Jensen; (right) Photo © Olson Photographic, Design: Northeast Cabinet Design, Ridgefield, CT

p. 193: Photos © Eric Roth

CHAPTER 7

p. 194: Photo © Olson Photographic

p. 195: (top to bottom) Photo © Olson Photographic, Design: Jean Callan King Interiors, East Haddam, CT; Photo © Mark Samu; Anthony Parkinson/Redcover.com, Design: Architect Marble Fairbanks and designer Michael Seibert; Photo © Mark Lohman, Design: www.lesliefineinteriors.com

p. 196: (left) Ken Hayden/Redcover.com; (right) Ken Hayden/Redcover.com

p. 197: (left) Photo © Mark Samu, Design: Mosaik Design; (right) Paul Ryan/Redcover.com

p. 198: Photos © David Duncan Livingston

p. 199: (top & bottom left) Photos © David Duncan Livingston; (right) Photo © Mark Samu

p. 200: Photo © Olson Photographic, Design: Jean Callan King Interiors, East Haddam, CT

p. 201: (top left) Photo © David Duncan Livingston; (bottom left) Photo © Mark Samu; (bottom right) Photo © Jon Jensen

p. 202: (left) Anthony Parkinson/Redcover. com, Design: Architect Marble Fairbanks and designer Michael Seibert; (right) Ken Hayden/ Redcover.com

p. 203: (top left) Ed Reeve/Redcover.com, Design: Harte Brownlee & Associates Interior Design; (top right) Photo © Eric Roth; (bottom left) Photo © Mark Lohman

p. 204: (top) Photo © Jon Jensen; (bottom) Photo © David Duncan Livingston

p. 205: (left) Photo © Mark Samu; (right) Photo © Eric Roth, Design: VAS Construction, Wilton, CT

p. 206: (left) Photo © Olson Photographic, Design: Kitchen & Bath Design Consultants, West Hartford, CT; (right) Photo © Jon Jensen

p. 207: Photo © Olson Photographic, Design: Jack Rosen Custom Kitchens/Joe Currie Designs, Rockville, MD

p. 208: Photo © Mark Lohman, Design: www. lesliefineinteriors.com

p. 209: (top left) Photo © David Duncan Livingston; (bottom left & right) Photos © Jon Jensen

p. 210: (left) Photo © Olson Photographic, Design: Olson Development LLC, New Canaan, CT; (right) Photo © Olson Photographic, Design: Cole Harris Architects, Westport, CT

p. 211: Christopher Drake/Redcover.com